QUANTUM EVENTUALITY THEORY

Romancing the Holy Grail of Physics

by

Grahame Gordon Innes

Bloomington, IN  Milton Keynes, UK

authorHOUSE

AuthorHouse™
1663 Liberty Drive, Suite 200
Bloomington, IN 47403
www.authorhouse.com
Phone: 1-800-839-8640

AuthorHouse™ UK Ltd.
500 Avebury Boulevard
Central Milton Keynes, MK9 2BE
www.authorhouse.co.uk
Phone: 08001974150

First published by AuthorHouse 4/17/2006

ISBN: 1-4259-2966-4 (sc)

Printed in the United States of America
Bloomington, Indiana

This book is printed on acid-free paper.

WELCOME TO THE WORLD OF:

QUANTUM EVENTUALITY THEORY
(Romancing the Holy Grail of Physics)

CONTENTS

FOREWORD

My first impulse is to explain the purposes for which I originally embarked on this venture, back in 1980. As a callow teenager of only eighteen, my motivations were somewhat ambitious and grandiose. Wrap up all areas of reality into a single package, explain everything, solve all loose ends and impress the world. During 1984 the second version of Quantum Eventuality appeared, then the final draft in 1991, from which this revision has been prepared. At that stage my original motivations had matured; I no longer wished to set the scientific community ablaze. I simply wanted to find my own answers to everything, with or without approval of the community. And time travel was on the agenda.

So, to what extent – if any – does this fourth revision resemble the original concept and present the same findings? It still explores the way in which the fabric and structure of existence govern all universal phenomena. There is still a particular intention to uncover a 'fundamental rulebook' from which the cosmos is engineered, and to which all phenomena are obedient. Along the way I still investigate the interactions between quantum events, analysing scientifically the nature of time as each investigation proceeds. The results of these inter-linked aims are still applied to the resolution of discrepancies in and between quantum theory, relativity and other cosmology theories. And finally the conclusions will still be passed on to more abstract areas of concern; philosophy, behaviour, psychology, morality, perception and language.

If so much is broadly the same, then exactly what is different? The third revision saw an exponential increase in the quantity and quality of equations and formulae used to substantiate the various arguments. In this final version I will seek to expand or simplify the mathematics as necessary, to make a clearer presentation to the layman. Certain subjects, such as gravity, will receive a more thorough examination. I shall be taking into account recent discoveries in cosmology, as well as newer theoretical models like superstring theory. These are the technical enhancements. But the main difference has got to be the writing style. This work is a book, in which I hold a conversation with the reader, not merely a scientific paper presenting a series of findings or abstractions.

The idea central to this theory – the Quantum Eventuality model itself, with its unique speculations on time – remains the backbone of the work. It is an idea that was conceived in 1980 and has remained a consistent feature of the theory. But, perhaps the real seed of this book was planted when, at the age of seven, I opened the pages of my first astronomy book. Consequently my thoughts were turned heavenward to the greatest mystery of all: existence. That is the real purpose of this book; it always has been and it always will be. If we fail to turn our faculties to the contemplation of existence, then there is hardly any point in possessing such faculties. And if we do not love what exists then we love nothing at all, and so make ourselves barren.

I invite everyone to enjoy reading this book as much as I enjoyed writing it, and to discover the special reverence for existence, which makes life meaningful.

Grahame Gordon Innes, 01/03/2006

INTRODUCTION

There is a fatal stillness all around, if such a thing as 'all around' exists. It is T - 3 seconds and we are aware only of a true void with a temperature of absolute zero. There is absolutely nothing surrounding us, not even the 'Cosmic Background Radiation' that Penzias and Wilson discovered. We are embedded in the ultimate sensory deprivation chamber though, of course, we are not really there, because the particles of which we are composed are already somewhere else: they are inside a minute invisible super-condensed speck called the Mono-bloc. So, our witnessing of these events is purely hypothetical. Hypothetically, however, we are about to observe the single most spectacular incident that has ever occurred in the visible universe; its birth.

All the matter that is to become galaxies, stars, planets, nebulae and quasars, is condensed into this ridiculously small dot, practically one-dimensional and possessing a density so great not even light can escape to advise us of its existence. It is a dot that is billions of times denser than a black hole, and is trying to contain unbelievably powerful concentrations of energy. T - 2 seconds and, hidden within that speck, magnificent changes and events are taking place that will determine the future course of cosmic evolution, at least in this local portion of universe. At sub-nuclear levels the laws of physics are re-forming, but something even more spectacular is going on; the creation of a single macro particle; the omnion.

In stark contrast to the surrounding freeze of nothingness, the undetectable Mono-bloc is experiencing temperatures of around 10^{11} K° or higher. But before going any further with this little scenario – to the 'Big Bang' and its aftermath – let us rewind the picture. After all, we know quite a lot about events since the Mono-bloc's explosion, but little about the previous history of existence. What made the matter of our universe get crushed into such a small space in the first place? Could something trap all that mass in an irresistible gravity well - a force that only the Big Bang itself could overthrow? Is it the mysterious 'dark matter' that nobody can find, or a super-massive Black Hole?

The most straightforward explanation is a simple reversal of the post Big Bang explosion; the matter was being pulled in and crushed by its own gravitational power, at near light speeds, in a 'Big Crunch'. Such a condition would have produced excruciating temperatures and pressures, pulling atoms and then particles apart. The collapse of atomic structure would have permitted further constriction through volume, but ultimately the near-light velocities provided the raw power necessary to feed the collapse, whilst momentum delivered the matter to a mutual rendezvous. The events that followed involved straightforward physics, even if somewhat exotic physics, but the implications of this 'Big Crunch' are far more significant. It means that what we call the Universe is only part of something even grander and more magnificent.

Every future is born from a past and our Universe is no exception. The future awaits you….

PAPER ONE:
A Single Field Theory of Existence

Entrée

Paper One is concerned with the structure and essence of reality. It explores the principles of Quantum Eventuality in the Cosmic Single Field Continuum (CSFC), which is the sum of the 'External' (all that exists apart from the self) and the 'Internal' (the self).

All that has ever existed, exists or ever will exist, can be expressed as a simple statement of its Quantum Eventuality; its duration in time, occupation of volume and quantity of mass.

Units of measurement to be used in Paper One;

Unit of Quantum Eventuality: 1 Queventum = $1gmcm^3s$ = Q

Unit of Quantum Integer Mass: 1 Quinmass = Qi = Em

Unit of Quantum Periodicity: 1 Querion = Qe = 0.5(Ve + Ca)

Unit of Quantum Deterioration: 1 Quandet = Qa = $\dfrac{m}{L/pT}$ = $\dfrac{pTm}{L}$

Unit of Quantum Elasticity: 1 Quelas = Qs = Qi x Qe

General Definitions

CSFC – The CSF Continuum (sometimes referred to as the Cosminium) = External + Internal = The Macrocosm = Universal Infinity = C

Photo-cosmos - that which we call the Universe, accessible to our perception.
X-cosmos – that which exists beyond the photo-cosmos.
C = P+X = E+I

Q = Quantum Eventuality/Event definition of a Temporal Edifice
T = Temporal definition of a quantum event or Temporal Edifice
V = Spatial definition of a quantum event or Temporal Edifice
M = Mass definition of a quantum event or Temporal Edifice

Temporal Edifice = a defined construction of mass and volume in time.

Section A: Elemental Factors, Infinite Continuity

Conservation Principles of the CSF Continuum

1. Temporal Conservation

Time can neither be created nor destroyed. We need to remove ourselves from our quaint, pedestrian view of time in order to appreciate the implications of this most important principle. Time is not about arbitrarily defined, artificial units of measurement called seconds, minutes, hours, days, weeks, months, years, centuries or millennia. Time is about the endurance of continuity. Our notion of time tries to tie down this endurance to even packets of a measurable phenomenon that pays no attention to the density of events defined by our arbitrary temporal units. Furthermore, our notion of time has only a local significance; a second is as it is only because of its relationship to minutes, then hours and days.

Of all the temporal units that we use to define events around us, days and years are the only true expressions that possess any universal significance, and even then their significance is limited to only this planet. Elsewhere in the universe, where planets and satellites rotate on their own axes or complete orbits in a shorter or longer duration than Earth, our day or year is irrelevant. Even within our own solar system there is no other planetary body on which it would be possible to talk of days and years that equate to Earth's axial rotation or solar orbit. We would have to redefine our units of time measurement to suit that other world's axial rotation or orbital duration. And stars themselves create even more problems. A star has no night or day and does not depend upon an orbit around another celestial bauble. If stars revolve around anything at all it would only be other stars, either because they exist in a binary or ternary system, or more commonly encircle a galactic hub composed of many billions of stars. As for nebulae, quasars and galaxies, humanity's preoccupation with clock-watching is made even less significant. Not merely because such phenomena are even further away from our limited terms of reference, but because on such a grand macro-scale our little existence is the blink of an eye too quick to even register.

The idea too, that the next parcel of time is yet to be created is a false one. The universe will face the future because the current universal time unit will simply never end, and the past because the universal time unit never had to begin. Is there any point, then, in discussing concepts of linear time such as past, present and future? The answer is yes and no. Yes, because it helps us to catalogue and define our own existence and the existences of all that we perceive. No, because universal time lies largely beyond our perception, on a scale that exceeds our imagination and which experiences no limit in its operation. In the universe time is only meaningful as the endurance of continuity, and that is its true and only definition. Each specific point in time proliferates throughout the entire macrocosm instantly and simultaneously at the same rate. It has no specific value, no termination and no birth. And localised effects fit into it at their own rate, not the other way around.

What exactly is meant by the 'endurance of continuity'? Simply put it is the extent to which a state of being continues unchanged. This is not an easy condition to conceive; it demands that throughout the entire universe at all levels there is no variance from a particular structural condition. Not even the exchange of any particle or energy is permitted. In fact, such a situation could not exist for reasons that will surely reveal themselves after any application of thought. To so exist would require a zero value for time, forbidding the movement of clocks with which to measure time. There would be no reaction between anything, no perception, no progress of events and most significantly of all no actual existence itself.

Thus we are forced to completely redefine our acceptance of how time works. If an electron orbiting an atomic nucleus completes 7000 spherical orbits per second then the endurance of continuity for it is minuscule and, consequently, a second defines a huge packet of aggregated temporal progress, at least in terms of its orbital movement. But if a mountain only loses a cubic centimetre of rock per month to erosion, then its aggregated temporal progress, in terms of erosion, is very slow. Time depends not merely on what we are contemplating, but on which aspect of that thing we are contemplating.

We have already seen that time cannot have a zero value, because existence itself would cease, events simply becoming impossible to occur and be witnessed. In fact interaction of any sort would be impossible. But time also cannot have a negative value since to do so would be to create anti-events; the undoing of existence. It is not merely that any one, isolated unit of eventuality cannot have a negative or zero time value; the whole of eventuality, the very universe itself, does not permit negative or zero values. This is the very core of Temporal Conservation; that no time can be created or destroyed. Time is an infinite commodity with a positive nature that excludes the possibility of cosmic birth or death. The endurance of continuity, then, is more a measure of change than the passing of time; the CSF Continuum proceeds through an infinite number of changes. Every new parcel of time is in reality just a new phase in the precise character and structure of the CSF Continuum. It does not matter where change happens, it still calls forth a new phase, separated from the previous one by an 'isochron'; an imaginary line between one temporal state of the universe and another.

We return to our consideration of the arbitrary nature of the unit lengths we measure time with. Supposing that we define our second as the time in which an electron passes across a quartz crystal of particular size and density. Our assumption is that electrons all behave in a uniform manner, that quartz crystals have no variance and that the force with which we move the electron is constant. Such an assumption is unjustified for two reasons. Firstly, even minute differences in one or all of the contributing elements to our measurement system could cause a change in behaviour. Secondly, even if the elements remain constant, a change in the laws of physics could produce quite different results. In the first scenario we may be unable to detect the change, even with the most sensitive equipment. In the second scenario, whatever we use to observe our measurement system will also obey the change in the laws of physics and thus no change will be observed. All we can hope to do, then, is to accept that measuring time is at best an artificial procedure and at worst a subjectively inaccurate one.

2. Linear Conservation

Time may be infinite and thus not truly measurable, but it is linear; causality is a one way sequence that is more than just an argument in logic. To say that event 'a' must precede 'b', otherwise 'b' would not happen, seems reasonable enough. But we must remember that 'a' may have several possible outcomes of which 'b' is only one, so it is not necessarily inevitable that 'b' follows on from 'a'. It may be just as likely that 'c' or 'd' are the outcomes of 'a'. But what is true and unquestionable is that, whatever outcome results from 'a' it is 'a' that is the instigation of the event sequence, and 'b', 'c' or 'd' that is the termination. Without the star there would be no nebula, but some stars have insufficient mass to create a nebula. Either way, nebula or no nebula, the star is a prerequisite. An earlier, different nebula may have been the interstellar nursery that helped create the star, but that is not the same phenomenon as the star's own end-of-life 'nebula' product.

Linear Conservation is a necessary consequence of Temporal Conservation; just as it is not possible for time to exist in negative or zero quantities, so it is not possible for the linear process of time to exist as a zero or negative quantity. This has profound implications for the temporal behaviour of the CSF Continuum. With the linear process of time always having a positive value, events must proceed in an onward direction step by step, each step being the consequence of its predecessor. Only the velocity with which the CSF Continuum unfolds can change, but it must always unfold forward. Cause and effect is not merely a logical convenience; it is the way the universe works.

3. Sequential Conservation

The laws of Temporal Conservation and Linear Conservation give rise to this third, associated law. Basically it states that if an event sequence does not follow one path it must follow another. There can never be a circumstance in which an event sequence follows a zero path or multiple paths (branching into two or more simultaneous progressions). This does not mean that a particular event, once started has only one possible outcome. It means that only one outcome can be realised. The instigation of an event sequence – 'a' – may lead to 'b' or 'c' or 'd'. But it does not lead to more than one of these possible terminators. As soon as the event sequence 'ab' unfolds it eliminates 'ac' or 'ad'. We can illustrate this quite easily.

A comet is on collision course with a planet; this is our instigation 'a'. If the sequence unfolds that the comet's course is not sufficiently deflected then it will hit the planet. But if its course is sufficiently deflected then it will not. There can be no situation in which the comet is sufficiently deflected yet still hits the planet, or is not deflected but does not hit.

Note; in saying 'deflected' I include any form of dissipation that would render the comet an inconsequential or vaporised mass before the impact point had been reached. I also point out that when I say an event sequence can only realise one outcome that is not the same thing as saying that there will only be one observed consequence resulting from the initial event. We can summarise these clarifications with an example. The Second World War had two possible outcomes; the allies won, or the allies lost: 'ab' or 'ac' but not 'ab/c'. In each case, each single outcome, there would have been numerous consequences flowing from the outcome. An outcome of an event necessarily includes all consequences as part of the outcome package.

4. Entropic Conservation

Entropy is the tendency of organised structures to dissolve over time. Decay is another way of putting it; living organisms die then rot, cars rust, buildings crumble and stars blow up in catastrophic self-annihilations. Since organised structures and systems carry more information and stored energy than chaotic ones, it is also said that entropy is the tendency for information and stored energy to be lost from a system over time. Indeed, since entropy is observable over time, it follows that the process of entropy must be connected to the first three Conservation laws above.

At this point I introduce you to the concept of 'temporal edifices', which will be discussed in greater detail later on. A temporal edifice is a 'structure in time'. This is more than merely the concept of physical structure enduring for a prescribed number of hours; it alludes to the object's temporal properties as the prime force behind its existence. This harks back to the 'isochron' mentioned in the first law - Temporal Conservation. There we talked about parcels of time, phases in the precise character and structure of the CSF Continuum. But that referred to temporal junctures omnipresent across the entire CSF Continuum. True, the ultimate temporal edifice would be the universe itself, and thus it would display 'isochronic' properties. But the real crux of 'temporal edifice' theory is an infinite number of self-contained finite structures with clearly defined inceptions and terminations.

It is the entropy of these 'temporal edifices' that is conserved through a mechanism of transmission by entropic osmosis. The dividing lines here are not so much 'isochrons' but 'isofluxes'. 'Isochrons' shadow all 'isofluxes' - the CSF Continuum experiences an 'isochron' line each time a part of reality develops an 'isoflux' line. Even if only one particle in the entire CSF Continuum underwent a phase transition across an 'isoflux', the CSF Continuum would still witness an 'isochron' at its own level. This is because every change to a finite part of the CSF Continuum is still a change in the precise state of the entire Continuum. In an infinity there will be an infinity of such changes, and so an infinite number of isofluxes and isochrons.

Entropy, as the tendency of organised systems to unravel, is a negative force, but what of its counterbalance; the creation of organised systems out of chaos? This is inverse entropy or Conjunction Rule, in stark contrast to regular negative entropy or Dissolution Rule. But how do these two principles operate alongside one another in an infinite cosmos? They operate by a process of balance, or Equilibration Rule. So, what are their values and behaviours?
The answer is as elegant as it is simple; in an infinite cosmos there are infinite changes and so entropy and its counterbalance can never be either zero or infinite themselves. There is no temporal edifice that can maintain existence whilst obeying zero or infinite entropy or inverse entropy. Across the macrocosm there exists an infinite amount of information which can therefore neither be added to nor subtracted from. So the entropic state of the macrocosm is a precise balance between those temporal edifices that are losing integrity and those gaining it. It is a continual trade-off as information lost from A is acquired by B in a never-ending cycle.

5. Conservation of Principle Obedience

If a temporal edifice does not submit to one set of physical principles, then it must submit to another set. No temporal edifice can come into being that submits to no principles or infinite principles, because existence itself requires well-defined rules. Furthermore the number of rules, to which even the CSF Continuum submits, must be finite, since an infinite quantity of rules would effectively tie reality up in an infinite set of contradictory principles.

On the face of it this may seem a conservation principle of little significance, especially following in the wake of four mighty temporal laws, but actually the Conservation of Principle Obedience is fundamental to the smooth running and stability of the CSF Continuum. Not only does it determine a stable framework governing the application of all other laws, but provides the very continuity that temporal evolution depends upon. In a way it is the CSF Continuum equivalent of a pituitary gland, regulating all other conservation functions.

6. Data and Pattern Conservation

So far we have examined the conservation laws that express the behaviour of time, entropy and cosmic law. Time is infinite and immeasurable, cause always precedes effect, cause leads to only one outcome, entropy is in equilibrium with its counterbalance and a temporal edifice can only obey one set of physical principles. Clearly these edifices contain structure and information, energy tied up and organised into meaningful patterns. On the face of it, after a period of decay, information may appear to have completely vanished from a system; the greater the interval between two observation points, the more 'vanished' the information appears to be. Alternatively the system may have encountered some disruptive force significant enough to have made a quantity of information vanish quickly.

The reality, however, is that no matter how much information has dissipated, no matter how thoroughly it has degraded, it had to go somewhere and therefore any pattern it formed or helped to form must also have followed it either in whole or in part. This is the principle of Conservation of Data and Pattern and it teaches us to apply scientific and forensic reasoning in our search for clues about the previous condition of the CSF Continuum or any of its temporal edifices. Palaeontologists have a clear picture of the progress of evolution, cosmologists talk meaningfully about events that happened billions of years ago, and police scientists determine the facts of a crime committed, who did it and even precisely when and how they did it.

7. Spatial Conservation

Between the nuclei and electron shells of atoms there is a huge void. Huge in atomic terms, that is; the particles themselves occupy far less space than the void yet to us even the whole atom is minute to a level that strains the imagination. If 85% of an atom is just empty space, then it follows that all matter could be crushed at the atomic level to only 15% of its original volume. Fusing some particles together would further reduce the size of matter, and dispensing with any 'empty' space between atoms can release even more volume from occupation.

But these are cosmetic changes, despite the profound effect they would have; they represent a more efficient use of space, just as exploding matter represents a waste of space. Therefore, in speculating on the conservation of space what we are really asking is can any volume be removed from or added to the CSF Continuum? Since the cosmos is an infinity the answer is that space can neither be created nor destroyed, and thus never carries a zero or negative value. Furthermore there can be no boundary or edge to the universe and there can be no microcosmic frontier at which objects might exist without taking up some volume. Infinite volume, a property of the CSF Continuum just as much as infinite time is, cannot belong to any finite part of the universe, so that all finite parts have boundaries.

It is a difficult concept to grapple with; a universe with no boundary containing temporal edifices with well-defined spatial limits. Yet we must remember that there are an infinite number of temporal edifices. A straightforward illustration may help: imagine that the cosmos is divided into a three-dimensional cubic grid. How big the cubes are is irrelevant; they could be cubic millimetres or cubic parsecs, it makes no difference. Into each cube we place just one electron mass. We have an infinite number of cubes containing an infinite number of electron masses, so our conservation laws are obeyed, yet the universe is still mostly empty space. If we then try to bring all the electrons together, without any space between them, we find that we can go just so far because we ourselves do not have an eternity in which to gather infinite electrons. Even if we did, our task would never be complete and it is likely we would end it voluntarily from ennui.

In our imaginary electron gathering challenge the best we could hope to achieve would be the cramming of as many electrons as possible into the least space. This brings us to the brink of theoretical disaster again; we would need infinite power in order to force so many negatively charged particles to co-exist in close proximity. Even if we replace our electrons with neutrons and permit gravity to assist us, the force we would need to exert on a galaxy-sized mass of neutrons would be pretty much galaxy-sized itself. Clearly our minute little structures, even if reconverted into pure energy, could never hope to pull the task off. So, the question now presents itself, where did the Mono-bloc come from and how did it get that way?

All the matter of the photo-cosmos crushed into a minuscule dot seems like an impossible construction, but the answer is as elegant as it is revolutionary. Firstly, Spatial Conservation is maintained because all that has happened is a quantity of matter being evacuated from all but one of the imaginary cubes in our immediately accessible 3-D grid. Around the Mono-bloc space has been sucked dry. Secondly, equilibrium is maintained because a micro space contains huge mass whilst a huge space contains micro mass. The whole situation is a case of reciprocal density, temporarily polarised in the most dramatic way, and possessing enough instability to create the spectacle of the 'Big Bang'. Although the problem of how to crush a photo-cosmos full of mass into an unstable dot is still a difficult one, the concept really only falters when we reach sub-atomic level. Discussions of that nature will be considered later, but for now we have established that Spatial Conservation works - and is indeed still a requirement - even during such exotic situations as the 'Big Bang'.

8. Conservation of Force and Exertion

Any quantity of mass that exerts force, or generates work, always causes a change within itself as well as within its environment. Since work cannot escape a closed system (the universe) it must therefore be redefined and redistributed for absorption into a new temporal edifice. This transfer of force from one temporal edifice to another is one directional and osmotic. The distinction between this conservation principle and the conservation of energy is a matter of measurement; conservation of energy specifically considers matter/energy exchange typically between two systems in balance. Conservation of force is only interested in the way in which change in one system is translated to change in another, even where systems are not necessarily equilibrated.

The overall state of force and exertion within the CSF Continuum is one of equilibrium, in parallel to the overall state of matter and energy. The difference is that it is often more difficult to measure gradual changes of force and exertion than the gradual exchanges of energy or matter. What is indisputable, though, is that for every quantity of force exerted by one temporal edifice, an identical force is received by another temporal edifice and the exchange of such force brings about alterations in both the originator and receiver.

9. Inertial Conservation (Angular Momentum)

Inertia is a generally understood principle; the tendency of mass to remain in a particular state of motion, until a force acts upon it to impart or absorb inertia. Thus the Earth continues to spin on its own axis and revolve around the sun. The huge stone remains still until a sufficient force is able to move it. Inertial Conservation shows that a mass either at rest or in motion cannot simply gain or lose momentum; it can only acquire it from or transfer it to another mass, so that conservation is maintained across both masses.

10. Centrifugal Conservation

A mass orbiting another mass will continue to do so whilst the centrifugal force acting upon it is directly proportional to the force binding both masses. If the binding force increases or the distance between the two masses decreases, then the satellite will move faster to conserve centrifuge. If the binding force decreases or the distance between both masses increases, then the satellite will slow down to preserve centrifuge. There are no negative or infinite values possible for this rule.

11. Magnetic & Gravitational Conservation

The magnetic and gravitational hold of two masses over one another is directly proportional to the quantity of both masses and the distance between them. Increasing mass or decreasing distance will result in a strengthening of magnetic or gravitational force. Decreasing mass or increasing distance will cause weakening of magnetic or gravitational force.

12. Conservation of Electric Charge

Any alteration in the electrical charge of a system must result in an equal and opposite alteration outside the system.

13. Conservation of Mass and Energy

Matter and energy cannot be created or destroyed, only converted in form, re-organised or transferred between one temporal edifice and another. Even the unison of matter with antimatter results in the transformation and liberation of both into energy, not true annihilation at all. No temporal edifice that exists can possess zero or negative mass, and no finite temporal edifice can possess infinite mass. Infinite mass is a property only of the CSF Continuum. This is also encapsulated in the first law of thermodynamics.

14. Thermal Conservation

Any transfer of thermal value in or out of a system is balanced by an equal and opposite transfer beyond the system.

15. Conservation of Quantum Eventuality

In an infinite cosmos nothing can be added or taken away, only altered from one state to another. We have already seen the conservation laws applying to temporal, spatial and mass factors of the CSF Continuum, and it is a logical consequence that these three factors of eventuality must obey a conservation principle collectively as well as individually. Since the CSF Continuum itself can neither acquire nor lose any time, volume or mass itself – because it is an infinite expression of these factors- it follows that Quantum Eventuality can not be acquired or lost by the CSF Continuum either. Quantum eventuality, then, is a product of time, volume and mass.

However, the quantities of time, volume and mass that exist in any of the infinite number of finite temporal edifices within the CSF Continuum can change. We have seen in the conservation laws of time, space and mass that these factors can be exchanged between different finite temporal edifices as long as the overall values of all involved edifices remain constant. It is therefore clear that eventuality can similarly be exchanged, according to its own conservation law. Eventuality can be removed from one system but it will always be transferred to another system or systems.

The Law of Conservation of Quantum Eventuality also permits changes within closed systems according to a precise set of rules. A change in one factor must be compensated for by an equal and opposite change in one or both of the other factors: a change in two factors must be compensated for by an equal and opposite change in the remaining factor. The possibilities are obvious;

1. Temporal extension / Spatial reduction
2. Temporal extension / Mass reduction
3. Temporal extension / Spatial & Mass reduction
4. Temporal reduction / Spatial extension
5. Temporal reduction / Mass extension
6. Temporal reduction / Spatial & Mass extension
7. Spatial extension / Mass reduction
8. Spatial extension / Temporal & Mass reduction
9. Spatial reduction / Mass extension
10. Spatial reduction / Temporal & Mass extension
11. Mass extension / Temporal & Spatial reduction
12. Mass reduction / Temporal & Spatial extension

The implications of this are astounding. In any closed system an alteration of one factor of eventuality leads to an opposite alteration in one or both of the remaining factors. Consider the closed system of the Mono-bloc; its explosion lead to massive spatial extension, accompanied by complementary temporal and mass reduction. Similarly, when a star goes nova it instantaneously loses mass and life span but acquires volume. Theoretically one could acquire mass by reducing volume or time or both. This is akin to what happens at near light velocity, as predicted in Einstein's relativity theory. It should be noted, however, that such events are quite exotic even in cosmic terms. Furthermore, once past the exotic phase, the systems involved settle all quantum accounts. The Mono-bloc continued its expansion only by irradiating mass into the surrounding volume it acquired, and losing time through entropic decay. The stellar nova, having blown off a huge swathe of its original mass, shrinks but settles into a less violent entropic state.

These events may have seemed like examples of closed systems defying conservation principles, yet even the relativistic journey returns the travelling matter to its original condition once the relativistic velocity is lost. More profoundly, though, what the quantum eventuality conservation law does tell us is most profound. The visible universe, or photo-cosmos, is not all that there is. Nor is any matter achieving near light velocity, any more than the nova is. They are all closed systems; self-contained and ultimately obedient elements in cosmic equilibrium.

Conclusions

By definition the universe is the sum of all possible events and phenomena, the product of temporal, spatial and material infinity. It is non-divisible and boundless, thus measurement is arbitrary and in real terms meaningless. Measurement can only be carried out on local, finite eventuality, where the factor being measured can be compared against the same factor in another finite eventuality. Therefore our observable universe is the photo-cosmos, a part or section of the macrocosm. Beyond what is observable lies the unknowable mass of the X-cosmos. The macrocosm is also the CSF Continuum, a cosmic single field continuum that consists of the External and Internal.

There must be direct relationships between all elemental factors of infinity, however they operate or are observed to operate, so that any two systems in equilibrium with a third are in equilibrium with each other. This is also known as the third law of thermodynamics.

Entropy in established systems or temporal edifices is negative and cumulative over time. It is intrinsic to all that exists and becomes observable as the endurance of continuity decreases. As entropy unfolds information is released from the temporal edifice, thus reducing comprehensible pattern. The CSF Continuum achieves equilibrium with entropic osmosis by isolated inverse entropy or Conjunction Rule. Local formations of organised eventuality arise from accumulated entropic by-products but, once stabilised, the new temporal edifices undergo their own entropy.

Larger aggregate pools of information yield more complex patterns, at the expensive opportunity cost of less comprehension per unit of information. But bear in mind that the CSF Continuum does not reveal limitless information, which is in constant flux anyway, to any finite portions of eventuality. This allows a loophole for the Internal to only comprehend some of the External. Even if the Internal could possess the infinity of information and pattern, comprehension would be reduced to zero. In real terms then, even billions of Internals, pooling their collective comprehension of the CSF Continuum, could never hope to possess infinite information, infinite comprehension of each unit of information or infinite comprehension of information patterns. As the pool of knowledge swells the potential for fully understanding each particle of knowledge decreases, and so does the potential for developing comprehension of the patterns involved.

We see this in practical terms as an ability to perceive local patterns such as galactic super-clusters, but without being able to know all things about those super-clusters. On the other hand we can comprehend individual elements such as the ionisation mechanism of gas molecules in the Kennelly-Heaviside layer, without being aware of the precise ionisation patterns across the whole atmosphere. The CSF Continuum information pool varies as local quantum event configurations cross isofluxes whilst the cosmos itself crosses isochrons. That it is thus never possible to know or understand all things is a subtle reflection of the Heisenberg Uncertainty Principle.

Naturally it also follows that the precise range of information, degree of comprehension and strength of pattern available to each Internal will vary considerably. In possessing different pools of information, comprehension and pattern recognition capabilities, the Internals acquire different outlooks from one another. It is therefore imperative that Internals seek to share information to achieve the broadest possible chance of perceiving the CSF Continuum accurately.

We are now progressing toward a valid and logical model for existence, a way of evaluating and measuring eventuality. It is almost a quest for the unknowable for, in illuminating the nature of an infinite cosmos, we will start with the packets or quanta that go to make up the CSF Continuum. Our three fundamental components of eventuality are time, volume and mass. If any of these factors did not exist (possessed a zero or negative value) then the CSF Continuum itself would not exist. There are four main reasons for this.

1. Time permits the construction of temporal edifices, the endurance of continuity that makes extension of mass through dimension possible. Without the temporal quality of the cosmos, mass would have no endurance in which to occupy, traverse and experience dimension. It is an elegant temporal/spatial symmetry, startling in the beauty of its simplicity yet mathematically profound. Temporal absence literally means that the fabric of the cosmos remains collapsed in nothingness.
2. Volume permits manifestation of mass in a stable framework that creates a reaction medium; it is the arena in which temporal edifices interact. Without space not even temporal infinity could hold mass, for mass without space would have nowhere to go, do or be.
3. To complete the picture of quantum eventuality, mass itself must exist. A cosmos in which there is temporal infinity and spatial infinity but no mass is still one without existence or eventuality. Time and space alone have no interaction and no fabric; they need mass to fuel the fires of eventuality within their framework, otherwise their framework is redundant.
4. Quantum Eventuality has a positive value in excess of zero. It is the product of its factors – time, volume and mass – and so each of those factors must have a positive value in excess of zero, in order for Quantum Eventuality to exist. $T \times V \times M = Q$. To set any of these factors with zero value invalidates the others; to set any with a negative value also invalidates the others. Since the value of Q for the cosmos is

infinity, and infinity cannot be divided, time, volume and mass for the cosmos must also be infinity. Q = TVM.

Our structure for the Universe, then, is as follows;

CSF Continuum/Macrocosm {C}

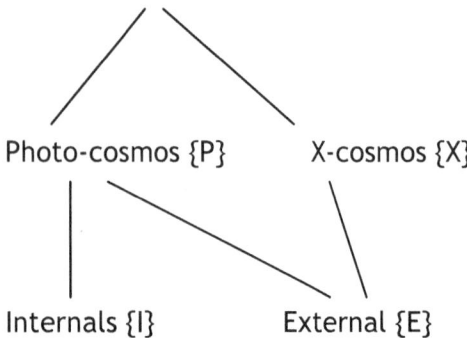

Photo-cosmos {P}   X-cosmos {X}

Internals {I}   External {E}

From the mathematical point of view we can say;

C = P+X = I+E
X = C-P = E-(P-I)
P = C-X = E-(X-I)
I = C-E = (P+X)-E
E = C-I = (P+X)-I
Q = TVM = C/0

$$T_c = \frac{X+P}{VM} = \frac{E+I}{VM}$$

$$V_c = \frac{X+P}{TM} = \frac{E+I}{TM}$$

$$M_c = \frac{X+P}{TV} = \frac{E+I}{TV}$$

Since the Photo-cosmos is finite, its separation from the Macrocosm cannot leave anything less than infinity, meaning that the X-cosmos is also infinite. Similarly the separation of the finite Internal from the infinite cosmos cannot leave anything less than infinity, so that the External is also infinite. How can finite quantities be removed from or added to infinity and the result is still infinity?

It is simply that the process of addition or subtraction is a false one, even with numbers. If you have a quantity 'a' from which you deduct 'b' the action is one of separation not destruction. 'b' still exists but separated from the rest of 'a'.
a - b = c does not require 'b' to disappear. Furthermore the very concept of numbers is false; 5 means nothing on its own, it only has significance if there are 5 stars or 5 planets or 5 litres of water. So, in our example the Internal is not actually removed from the CSF Continuum, it is just being considered separately. The photo-cosmos similarly cannot be 'deducted' from the macrocosm, just thought of on its own.

This realisation holds true no matter how many finite quantities or how much finite material you wish to consider separately from the infinite macrocosm; the remainder is always still infinity because the reality of subtraction is merely to consider a quantity of something apart from the rest. For the same reason you cannot add; to start with 3 pebbles and 'add' 2 does not require creation, only the drawing of different boundaries. The 2 'new' pebbles were somewhere else in the universe, now we are thinking of them as being with the other 3. The act of addition does not even demand that we physically move anything; it is always merely a way of thinking, an exercise in how we are choosing to perceive.

Does a similar truth exist for multiplication and division?

In the table above $M = \dfrac{X+P}{TV} = \dfrac{E+I}{TV}$

What this equation tells us is that the total mass of the cosmos is equal to the {eventuality of the X-cosmos and photo-cosmos}, divided by {total time multiplied by total volume} or the sum of External and Internal divided by the same factor. Multiplication is almost a return to our earlier definition of time; the endurance of continuity. To multiply a thing is to say that there is not merely what is immediately perceived, but a greater proliferation. But it is always subject to upper and lower limits; to multiply infinity will always yield infinity and to multiply zero will always yield zero. Division of these upper and lower limits will always yield the same results as well. In considering the universe, multiplication and division are as without meaning as addition and subtraction. Only finite quantities can undergo such pedestrian calculus and even then the process is still artificial. Division, in essence, is an antidote to the endurance of continuity; discontinuity.

If I have 5 atoms and multiply them by 5, I now have 25 atoms; the additional 20 atoms have had to come from somewhere. They have had to be deducted from other matter, which is now 20 atoms light. Conversely, if I start with 100 atoms and divide them by 5, I leave myself with only 20; but 80 atoms have not simply ceased to exist. They have been added to other matter, which is now 80 atoms proud.

This method of abstract cosmic accounting might seem a diversion but it has colossal significance in the Quantum Eventuality model of the universe. Firstly it sums up the very nature of the infinite; the CSF Continuum has to balance, has to obey all the conservation laws in order to exist. Secondly these thoughts, about the way that arithmetical operations in an infinite universe are fallacious, help to liberate our perception so that an understanding of Quantum Eventuality is possible.

In the equation table above we saw $Q = TVM$.

This is our starting point: Eventuality is equal to time by volume by mass, and for the CSF Continuum each factor has an infinite value. It is a measurement of existence, not the creation of something from nothing. But what does it really signify?

At the beginning of this section I discussed the primary role of time; this was no exaggeration. Of the three factors it is the most fundamental of all. It is almost independent of the others, since - in taking up a quantity of time - a finite amount of eventuality does not borrow any time from anywhere else. All events have at their disposal the temporal medium, which is omnipresent and simultaneously proliferated throughout the cosmos. Of space and mass the same is not true at all. In order to acquire volume an eventuality must occupy a space that becomes unavailable to other eventualities. And acquiring mass likewise removes a quantity of that factor from temporal edifices already in existence. Time, the endurance of continuity, really is universal and is the start of all things.

The universe, even if possessing only temporal infinity, would still itself be endless. And, as infinity, it has no beginning or end, no boundary and no edge. It cannot be added to or taken away from. Nor can it be multiplied into anything greater, nor divided into anything lesser. It is unassailable.

Remember that in constructing a temporal edifice - a reasonably independent accumulation of quantum eventuality - we do not need to deprive anything else of time. But mass and volume do need to be reassigned from elsewhere. As our earlier exploration revealed, subtraction is separation not destruction. And thus addition is unification not creation. We are forced to concede an important point of philosophy and semantics, on our way to understanding what could be called the Quantum Weirdness of reality. In the infinity that is the cosmos, there can be no creation or destruction, only redistribution. Therefore our concepts of creation and annihilation are as false as our concepts of arithmetical operation. The truth is that the cosmos is always changing, in an endless cycle of cosmic births and deaths. But these are merely the transitions of one set of temporal edifices, across isochrons and isofluxes, to a newer set of temporal edifices. New made from old - it is recycling at its most impressive.

Although Paper One is primarily concerned with the structure and essence of reality, we cannot ignore basic facts. Fiction and imagination are a part of reality. The Internal is a part of the External. The Photo-cosmos is part of the macrocosm. Our notion of separateness is almost as inaccurate as that of creation and destruction. But even more importantly, in our relationship with the CSF Continuum, are the elements of information, pattern, comprehension and entropy.

Let;

Ω = Total information in a system

Δ = Pattern strength

μ = Comprehension of each unit of information

$\emptyset$ = Information lost to Entropy per time unit

$\in$ = Entropy

T = number of time units

Then $\Omega\mu$ = total comprehension and,

$$\Delta = \Omega\mu - \phi T$$
$$\Omega\mu = \Delta + \phi T$$

$$\Omega = \frac{\Delta + \phi T}{\mu}$$

$$\mu = \frac{\Delta + \phi T}{\Omega}$$

As $\Omega\mu$ increases so does Δ, however, it should be noted that the larger a body of information becomes, the less strong is awareness of each element of information. Since $\Omega_c = \Sigma$ the total awareness of each element of information (for the entire CSF Continuum) is zero. Consciousness, then, must content itself with contemplating a finite quantity of information at finite pattern strength, in order to possess a sufficient grasp over the elements of that information. We can present this as a graph;

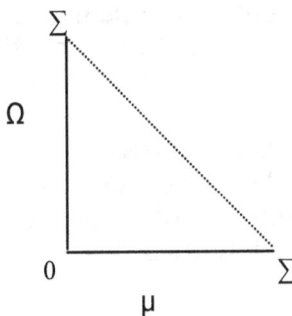

What does this relationship teach us about our ability to understand the universe? As the total information in a system approaches infinity, potential comprehension of each unit reduces to zero. Therefore, the more information we absorb from the universe, the less we are aware of each individual piece of information. But with greater information comes a clearer understanding of pattern; knowing more about something helps us comprehend the pattern of its existence, but makes it harder to be aware of the smallest quanta of information in full. Alternatively, to concentrate on individual bits of information removes our vision from the larger picture; we cannot fully know both the quanta and the amalgam of quanta. We cannot see the pattern if we are too busy contemplating a small portion of it. This fits in with our model of entropy, which can be neither infinite nor zero for finite phenomena.

We can also construct a graph to depict this relationship;

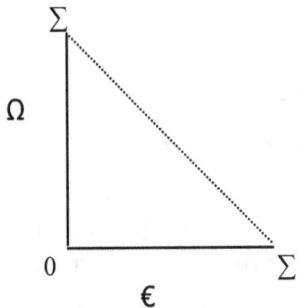

From this we see that as information increases to infinity, entropy decreases to zero. The relationship is exactly the same as that between information and comprehension.

Can it really be true that awareness of elements from which a temporal edifice is constructed, is identical to the entropy of that edifice? Why not? After all, we know that one definition of entropy is loss of information over time. And if we understand part of something is it not feasible that the only reason we understand that part is because it represents information lost through entropy?

This principle is at its clearest in considering the CSF Continuum; information is infinite and therefore entropy is zero. Our understanding of each element is also zero because no information is being lost and so we as observers cannot gain any information. In everyday life, though, we do not observe the infinite and it becomes possible to acquire information as a result of entropy. Assuming the same fixed quantity of lost information for all things, the larger a temporal edifice is the longer it will take to decompose. It releases the same quantity of information per second as a smaller edifice, under this imaginary scheme. Thus the information strength yield per second is $\emptyset/\Omega$ and for a period of several seconds $T\emptyset/\Omega$. Clearly the figure for comprehension will be higher for a smaller source, where both sources lose the same quantity of information across the same time period.

In a situation where nothing exists, entropy stands at infinity whilst information is zero. No information can be lost because there is no information, and so understanding of each element stands at infinity too, for the understanding of zero can be nothing other than infinite. But how does the universe really work in between infinity and zero; in the realm of everyday experience?

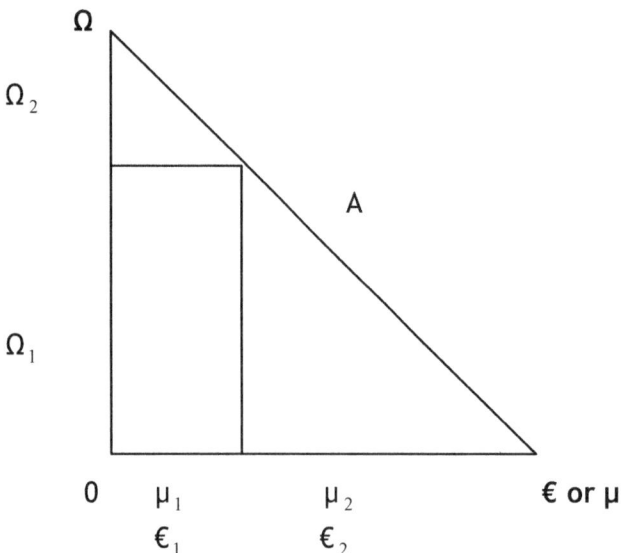

We already know that angles ΩA and $\mathbf{€}A$ or μA are each 45° so that $\Omega_1 = \mu_2$ or $\mathbf{€}_2$ whilst $\Omega_2 = \mu_1$ or $\mathbf{€}_1$ and therefore the larger a thing is the less understanding of it we can proportionately master. This is an easily observable general rule but the less obvious part of it is the implication for entropy. Along with quantum eventuality, information and pattern, it all comes down to the principle of equilibrium. Infinite values are conserved through infinite redistribution between an infinite number of finite parts. As far as the CSF Continuum is concerned entropy does not exist and so an unlimited amount of information enjoys an endless temporal existence. Decay is a finite manifestation of entropy and is balanced by isolated inverse entropy or Conjunction Rule through entropic osmosis.

A galaxy will outlive any one of its stars, and a star will outlive its planets. A life form has a greater life expectancy than its cells, and the larger it is the longer it will survive the normal routine of existence. Continual protection and sustaining refurbishment of a small mass may help to extend its integrity, whilst the introduction of disruptive forces may shorten the integrity of a large mass, but the clear principle is that (excluding intrusions) greater mass leads to greater endurance of continuity. As a point of interest, when an organism perishes it is usually the failure of a part of it that is responsible; the heart, brain, kidneys or liver cease functioning before the being they serve ceases functioning.

So, negative intrusions by external forces cause an increase in rate of decay leading to premature failure. Positive intrusions cause a reduction in rate of decay leading to post-mature failure.

Our graph allows us a remarkable insight into the trigonometry of the cosmos. With the hypotenuse angles both at 45° the universal Tangent (Sine θ / Cosine θ) becomes unity (1), whilst the oscillation or wave functions begin with basic values;

$$\text{Cosine } \theta = \text{Sine } \theta = \frac{\sqrt{2}}{2} = 0.70710678$$

(Where θ is our 45° angle at both extremities of the graph). It is hardly surprising that on the grandest macrocosmic level these factors of infinity should betray the trigonometric basis of all waves including the sub-atomic and photonic quanta of theoretical physics and quantum mechanics.

The general wave equation is $X = A \cos(\dot{\omega} \times t + d/\lambda)$

Where;
X = distance from equilibrium point
A = amplitude of motion

$\dot{\omega}$ = angular frequency of motion
t = time
d = spatial position on wave

λ = wavelength

These relationships of time, entropy, information and mass all relate to how much information a temporal edifice releases per time unit, in proportion to its total information.

And they lead us to the **Information / Entropy / Time Paradox**.
This states;
A: as entropy increases, information decreases
B: as time progresses, entropy increases
C: as time progresses, information increases
Such a paradox takes the straightforward shape of 'if A then B, If C then A, if C then not B'. But it is not a true paradox and is easily resolved by looking more clearly at each of its tenets and its general terminological interpretations. We must remember that the

tangential value of information, entropy and comprehension is '1', and therefore cosmic unity. Whatever statements are made, by whoever makes them, the cosmic tangent is still '1'. It will also help to redefine our viewpoint of cosmic interaction; the Internal is oneself and the External all that exists beyond oneself. Within the CSF Continuum are many billions of individual consciences; they are the plural Internals, and they still relate to the External (all that exists beyond the many selves).

Looking at our first statement (A), the information lost as entropy increases is transferred elsewhere, resulting in a lowering of entropy in the new location. For example, the Mono-bloc had a hidden, briefly negligible entropic value whilst losing virtually no information, but the space around it had infinite entropy and zero information. The 'Big Bang' allowed the Mono-bloc to begin releasing information in huge amounts whilst increasing its entropy exponentially. The counterbalance came as information-laden matter poured into the surrounding space where it began decreasing entropy rapidly. As time progressed, this matter reached a lowest entropic value and maximum information content. A critical watershed was reached at which point regular entropy / information loss began. As information leaves systems it builds up elsewhere over time, thus increasing in some places even as it decreases in others. It is merely a question of understanding where, when and why; of knowing how to understand the subtle trade of information that is a constant feature of an infinite cosmos.

The information lost in A is irradiated away from some parts of the CSF Continuum and collected by other parts, thus maintaining the Equilibration Rule. Only when consciousness intercepts this information over a period of time does it appear to increase. The cosmic background radiation is just such an example of lost information being intercepted by consciousness and embraced as an increase in a pool of information. A, B and C are all true, depending from which perspective one observes the External, and specifically which part of the External that perspective observation is directed at.

The final point to address in this artificial paradox is the strength of interception. Not all consciousness receives the same information at the same strength. On the surface the logical question arises, why? Should not the exact same information be just as easily grasped by one Internal as another? Of course it should, but unfortunately consciousness is even more asymmetrically distributed and 'lumpy' than the manifestations of Quantum Eventuality. Thus one being may immediately realise the significance of the phenomenon it has just witnessed, whilst another is not even aware of the phenomenon at all. This inequality in the strength of consciousness may be demonstrated by another graph.

No of bytes of information

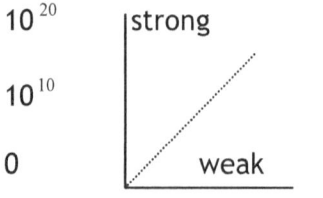

The dotted line itself is another 45° angle and represents mean comprehension level as a straightforward reciprocal of however much information is contained. It is a median between the unconscious computer that possesses an indefinite quantity of information but zero comprehension, and a theoretical mind that possesses infinite understanding of nothing.

This mean line with its 45° angle is also tangential and reflects an ideal comprehension to information ratio of '1' – unity. 10^{20} bytes of information multiplied by comprehension of 10^{-20} per byte = 1. In algebraic terms: $a^y \times a^{-y} = 1$.

Consciousness that is a strong receptor traces out a wide arc yield with an angle >45°; this is the realm of the 'generator mind' that gains an amazingly accurate picture of reality based on a paucity of information, as the great philosophers of ancient Greece did. Alternatively we are looking at the creative genius like Mozart who, with a minimum of training, wields elements of knowledge to create the astounding and elegant; a suitable reflection of the universe.

Consciousness that is a weak receptor traces out a narrow arc with an angle <45°; unable to build a sufficiently detailed or accurate picture of reality no matter how much information is placed before it, only rudimentary results are forthcoming even with guidance and tuition. This is the realm of those who mock and insult generator minds. It is the realm of the negative and acidic personality that sets out to ridicule and destroy order.

The Specific Interpretation of Eventuality

The definition of eventuality is 'manifestation of mass through space and time'. If any element of this definition is withheld or removed, eventuality ceases. Mass alone is not sufficient to prop up existence, because space is the only medium across which mass (matter and energy) can interact and time is the only medium through which interactions can become effective. Before we arrive at measurements of eventuality we must refine our definition further.

The interactions of mass require the following universal qualities;
1. Freedom of movement
2. Freedom of chemical and nuclear transformation
3. Freedom of temporal exchange

These can only be provided by spatial and temporal media, so that if space or time are absent from our equation of eventuality it becomes impossible to satisfy these requirements. Movement – both kinetic and dimensional – must be both present and free if mass is to react to stimulus and with other mass, and can only happen across spatial medium. Chemical and sub-nucleonic freedom involves space, without which chemical reactions and particle transformations are impossible. Whatever form mass or energy take, an inability to react, because space is absent, means that no existence is possible. Finally freedom of temporal exchange – the primary mechanism by which entropy and information operate – is vital to all functions of existence. No time means no exchange and no eventuality.

Space barren of mass and deprived of temporal medium also invalidates eventuality. If it were possible to create in deep space a vast bubble, which contained absolutely nothing, not even light or radiation, into which nothing would be able to pass, the result would be sterility. Even if time still worked within the bubble, nothing else would; so, no eventuality. Without time, not even the hypothetical bubble itself could exist. Space, it seems, also requires certain universal qualities;

1. Freedom to acquire and release mass
2. Freedom to merge boundlessly with other space
3. Freedom of temporal continuance

The final and most important element of eventuality is surprisingly dependent on the others, because time requires;

1. Freedom to construct all mass
2. Freedom to support all space
3. Freedom to be omnipresent

Eventuality – existence – is the product of the three fundamentals; temporal, spatial and material. Although time could theoretically stand alone it would have no purpose in so doing; it may be the most fundamental of the three but without the other two it is no nearer to completing the full picture of eventuality. Even with just two of the three fundamentals, eventuality refuses to happen.

We have already encountered the basic measurement of eventuality; the Queventum (Q), or $1gmcm^3s$. This is the mathematical and scientific definition of eventuality and it simply equates to the sum of mass operating across space through time. In the plural it is Queventa for any phenomenon under consideration. For the infinity of the cosmos our representation of all Quantum Eventuality is Q.

The basic unit of eventuality, the Queventum, will always be one gram of matter occupying one cubic centimetre for one second. We can tinker with these figures as much as we like but one Queventum has no greater or lesser significance than any other Queventum. For example;

0.5 grams of matter occupying 0.25 cm^3 for 8 seconds is still 1Q, as is 50 grams of matter occupying 2 cm^3 for 0.01 seconds. These are little more than basic transformations of Quantum Eventuality and hark back to our array of possible transformations within closed systems;

1. Temporal extension / Spatial reduction
2. Temporal extension / Mass reduction
3. Temporal extension / Spatial & Mass reduction
4. Temporal reduction / Spatial extension
5. Temporal reduction / Mass extension
6. Temporal reduction / Spatial & Mass extension
7. Spatial extension / Mass reduction
8. Spatial extension / Temporal & Mass reduction
9. Spatial reduction / Mass extension
10. Spatial reduction / Temporal & Mass extension
11. Mass extension / Temporal & Spatial reduction
12. Mass reduction / Temporal & Spatial extension

Such transformations seem at odds with the conservation laws, and it is hardly surprising that they require very exotic circumstances. In order to understand this principle let us consider two very different types of closed system; that which is finite and the CSF Continuum which is infinite. Ultimately we will see just how closely they are linked.

Our starting point for the finite is Q itself, so let us begin by assigning values to its factors. Say:
T = 1500 seconds
V = 4750 cm^3
M = 10800 gm
Q = 1500 x 4750 x 10800 = 7.695 x 10^{10} gmcm^{3}s

This may seem like a huge amount of eventuality, but in reality 10.8 kg taking up the equivalent of a cube of side 16.80987703 cm for only 25 minutes is really not much at all. Under normal circumstances, like any other Quantum Eventuality or Temporal Edifice, this one would react with matter, space and time around it. For this test of the nature of eventuality, however, it has been isolated from the rest of the CSF Continuum. Now it is subject to the 12 transformations in the table and these are entirely relativistic effects. That is to say that a relativistic effect is always produced where T, V or M change value but Q remains constant.
In our first set of relativistic transformations let us assume that whilst one factor increases only one of the other factors compensates. Our first array looks like this;

1. Temporal extension / Spatial reduction, V = Q/TM
2. Temporal extension / Mass reduction, M = Q/TV
3. Temporal reduction / Spatial extension, V = Q/TM
4. Temporal reduction / Mass extension, M = Q/TV
5. Spatial extension / Mass reduction, M = Q/TV
6. Spatial reduction / Mass extension, M = Q/TV

All of these calculations have the general form a = Q/bc, where 'a' is the compensating factor of eventuality and bc represents the remaining factors where change has occurred. Imagine in our example above that in two different operations we firstly increase one factor by 25%, then decrease it by 25%. Our table now looks like this;
SCENARIO 1

Ex	Original	+ 25%	- 25%	Compensation
1 / 3	T = 1500	T_2 = 1875	T_2 = 1125	$V_2 = Q/T_2 M$
2 / 4	T = 1500	T_2 = 1875	T_2 = 1125	$M_2 = Q/T_2 V$
5 / 6	V = 4750	V_2 = 5937.50	V_2 = 3562.50	$M_2 = Q/TV_2$
1 / 3	V = 4750	V_2 = 5937.50	V_2 = 3562.50	$T_2 = Q/V_2 M$
2 / 4	M = 10800	M_2 = 13500	M_2 = 8100	$T_2 = Q/VM_2$
5 / 6	M = 10800	M_2 = 13500	M_2 = 8100	$V_2 = Q/TM_2$

Thus our new values become;

Ex	+ 25%	a = Q/bc	- 25%	a = Q/bc
1 / 3	$T_2 = 1875$	$V_2 = 3800$	$T_2 = 1125$	$V_2 = 6333.33\,\dot{}$
2 / 4	$T_2 = 1875$	$M_2 = 8640$	$T_2 = 1125$	$M_2 = 14400$
5 / 6	$V_2 = 5937.50$	$M_2 = 8640$	$V_2 = 3562.50$	$M_2 = 14400$
1 / 3	$V_2 = 5937.50$	$T_2 = 1200$	$V_2 = 3562.50$	$T_2 = 2000$
2 / 4	$M_2 = 13500$	$T_2 = 1200$	$M_2 = 8100$	$T_2 = 2000$
5 / 6	$M_2 = 13500$	$V_2 = 3800$	$M_2 = 8100$	$V_2 = 6333.33\,\dot{}$

Clearly whichever factor changes, the factor that compensates does so by a different percentage. We can easily calculate the compensation degree;

$$F = \frac{1}{1 \pm p/100}$$

In this equation 'p' is the percentage increment or loss of the altered factor. Thus in adding 25% we obtain from this result; F = 1/1.25 = 0.80, indicating that we multiply the compensating factor by 0.80 or deduct 20%.

And in taking away 25% we obtain the result; F = 1/0.75 = 1.3333$\dot{}$, indicating we multiply the compensating factor by 1.3333$\dot{}$, or add 33.33$\dot{}$%. But what happens when two factors increase or decrease, and the third compensates?

We have a new, second array; **SCENARIO 2**

    7. Temporal extension / Spatial & Mass reductions, T = Q/VM
    8. Temporal reduction / Spatial & Mass extensions, T = Q/VM
    9. Spatial extension / Temporal & Mass reduction, V = Q/TM
    10. Spatial reduction / Temporal & Mass extension, V = Q/TM
    11. Mass extension / Temporal & Spatial reduction, M = Q/TV
    12. Mass reduction / Temporal & Spatial extension, M = Q/TV

Original TV = 7125000, TM = 16200000, VM = 51300000

Again we have the general form a = Q/bc but now we have to consider whether both the altered factors are altered by the same degree or by different degrees. If both factors are changed by the same percentage, 25% for example, then;

$$b_2 = \frac{b_1 c_2}{c_1}$$

$$c_2 = \frac{c_1 b_2}{b_1}$$

$$bc_1 = \frac{abc_2}{a_1}$$

$$bc_2 = \frac{abc_1}{a_2}$$

$$a_2 = \frac{abc_1}{bc_2} = \frac{a_1}{bc_2 / bc_1}$$

We have a new table of values;

Ex	+ 25%	a = Q/bc	- 25%	a = Q/bc
7 / 8	$V_2 = 5937.50$ $M_2 = 13500$	$T_2 = 960$	$V_2 = 3562.50$ $M_2 = 8100$	$T_2 = 2666.66^{\bullet}$
9 / 10	$T_2 = 1875$ $M_2 = 13500$	$V_2 = 3040$	$T_2 = 1125$ $M_2 = 8100$	$V_2 = 8444.44^{\bullet}$
11 / 12	$T_2 = 1875$ $V_2 = 5937.50$	$M_2 = 6912$	$T_2 = 1125$ $V_2 = 3562.50$	$M_2 = 19200$
TV	$11132812.50 \ cm^3 s$		$4007812.5 \ cm^3 s$	
TM	25312500 gms		9112500 gms	
VM	$80156250 \ gmcm^3$		$28856250 \ gmcm^3$	

Now let us consider changes in two factors by different percentages. The equations for the new value of 'a' hold true, but do those for b and c? They do not.

$$\% \text{ change in factor a} = (1 - \frac{bc_1}{bc_2}) \times 100 = 100 - \frac{100a_2}{a_1}$$

Let us say that one factor changes by 25% and another by 17.9% in the same direction, by how much will the third factor change in the opposite direction to compensate? Based on the example we have been using: **SCENARIO 3**

Original	+25%	+17.9%	a = Q/bc	Opposite %
T = 1500	$T_2 = 1875$	$M_2 = 12733.2$	$V_2 = 3223.0704$	32.14588
T = 1500	$T_2 = 1875$	$V_2 = 5600.25$	$M_2 = 7328.244$	32.14588
V = 4750	$V_2 = 5937.50$	$T_2 = 1768.5$	$M_2 = 7328.244$	32.14588
V = 4750	$V_2 = 5937.50$	$M_2 = 12733.2$	$T_2 = 1017.8117$	32.14588
M = 10800	$M_2 = 13500$	$T_2 = 1768.5$	$V_2 = 3223.0704$	32.14588
M = 10800	$M_2 = 13500$	$V_2 = 5600.25$	$T_2 = 1017.8117$	32.14588

Clearly, the relativistic equations for b and c are no longer applicable, although we can derive some generic formulae to cover all possible contingencies: for example in cases where $b_2 = b_1 \sqrt{\dfrac{c_2}{c_1}}$ then $c_2 = \left(\dfrac{c_1 b_2}{b_1}\right)^2$

Where 'a' is the single compensating factor, b and c alter individually or together by either the same or different percentages, Q is the sum of a x b x c and p is the percentage by which 'a' changes we can derive field equations based on ratios;

Ratio is a: b: c

p % of a = $\dfrac{100\,pa}{Q}$ % of Q = $\dfrac{100\,pa}{b}$ % of b = $\dfrac{100\,pa}{c}$ % of c = $\dfrac{100\,pa}{bc}$ % of bc

In considering these relativistic transformations of eventuality it must be emphasised that what we are dealing with are perturbations within closed systems. Changes in time, space and mass that do not involve any external interactions with the rest of the CSF Continuum. Such instances are extremely unlikely, since it is nearly impossible for any quantum eventuality to exist in the CSF Continuum yet not react with it. Given that it is in the nature of all quantum events, or temporal edifices, to react in some way with any other quantum events encountered, exactly how could such isolation come about?

The answer is that under normal circumstances isolation is not possible and so relativistic effects would not be encountered. However, Einstein revealed to us at least one relativistic transformation that can be arranged, tested and observed; acceleration to the velocity of light, or at least close to it. He predicted that time for an object travelling at or near light velocity would become distorted, having a counteractive effect on the object's volume and mass. To the rest of the universe such an object would appear to acquire exponential mass, whilst shedding length (one of the dimensional constituents of volume). It would also live a lot longer than matter moving around at more pedestrian speeds. The effect has been tested in particle accelerators like CERN, where short-lived exotic particles accelerated to light speed, or almost light speed, increase their lifespan. Their mass also increases, though it has not been possible (I believe) to ascertain the effect on their volume. It is not unreasonable, though, to deduce that their spatial factor must reduce in order to permit temporal and mass extension.

So far we have examined relativistic effects in closed systems, but have they really disobeyed the conservation laws? The short answer is no. No real mass, space or time has been lost or acquired. Which brings us to a conclusion that is truly staggering; within the closed system undergoing relativistic distortions, all three factors of eventuality are interchangeable! As peculiar as this may sound it is the inescapable deduction.

The well-expounded illustration of an astronaut on a near light speed (time dilation factor 0.00001181818182) voyage to the Andromeda Galaxy perfectly summarises relativistic transformation. He ages 26 years on his journey to Andromeda, and 26 years coming back; the rest of the universe undergoes 4,400,000 years of evolution compared to his total ageing of only 52 years. For the astronaut and his spacecraft time has extended; he perceives the universe speeded up to a rate of 23.5042735 hours for every second of his own time. In his year the universe ages 84.61538462 millennia. He also sees it shorten and thus become smaller; losing both time and space, the External logically would appear to gain mass if the astronaut could weigh it. From the External's viewpoint, his transformation is extension of time and mass but spatial reduction.

We can plot the shifts of both External and Internal on a simple grid:

Factor ::::::::::::::::: Astronaut (Internal) :: Universe (External)

Factor	Astronaut (Internal)	Universe (External)
Temporal (Longevity)	$+ r$	$- r$
Spatial (Dimension)	$- r^2$	$- r$
Material (Mass)	$+ r$	$+ r^2$

It is intriguing that from both perspectives the volume of the other divides and reduces exponentially. As a result of this, owing to the particular relationship of mass to volume (density) mass multiplies and increases exponentially. It is only the temporal factor that moves in opposite ways. In order to preserve total Quantum eventuality, the astronaut (and his spacecraft) will appear to lose volume by a factor of r^2 (where he gained both time and mass by a factor of r). On the other hand, he will see the universe lose both time and volume by a factor of r, whilst gaining mass by a factor of r^2. The total of these distortions is thus zero.
$$2r - r^2 - 2r + r^2 = 0.$$

It is a very elegant equilibrium and a poignant example of relativistic transformation, but does it really represent increases or decreases of Quantum Eventuality factors across the CSF Continuum, in direct contravention of the conservation laws? The answer is still no, not merely because overall Q has remained constant, but particularly because the conservation of entropy is preserved and the transformation is only a temporary effect anyway. Indeed, relativistic distortions of this nature can only ever be temporary, because the circumstances necessary to bring them about are not indefinitely supportable.

Given that no real gain or loss in values of TVM can occur across the CSF Continuum, and that we have already concluded that TVM are interchangeable during relativistic situations, the conservation laws remain unchallenged. Only the matter of entropy still needs to be considered in relation to these transformations, then, and I hasten to remind you that whatever affects entropy will be as temporary as the transformation causing the effect.

When a Quantum Event is artificially pushed into relativistic perturbation, it becomes closed. This means it achieves a heightened level of independence from the rest of the CSF Continuum. Normal entropy behaviour therefore no longer applies. The most important aspect of this entropic separation is the loss of entropic osmosis; the free trade of information between temporal edifices. Without such trade the position of any associated isofluxes shifts away from being between one edifice and the next adjacent edifice, and becomes a barrier between one edifice and the rest of the cosmos. Total separation is naturally not possible, but entropic isolation is the next best thing. Our fictitiously fast astronaut experiences a dramatic increase in the entropy of the

External, whilst the External perceives the astronaut to have virtually no entropy. Not surprisingly, the factor of increase or decrease is 84615.38462 in each case.

I have mentioned Quantum Weirdness previously: the tendency of eventuality to depart from Quantum Orthodoxy, under exotic conditions, or to appear contrary to expectation. Because this is not a rule, but a departure from rules, it is difficult to try and codify it like a rule. But the principle still holds that whenever reality departs from the expected, Quantum Weirdness is involved. Relativistic transformation is perhaps the strangest and most pronounced example of this phenomenon.

There are only two questions left to tackle in our specific interpretation of eventuality. Can relativistic transformation affect the entire cosmos at once, and does it have commonplace manifestations?

The first question is easy enough to answer. In the macrocosm, the three factors of eventuality all have an infinite value; they cannot be added to, subtracted from, multiplied or divided. So, across the entire cosmos relativistic transformations are not possible. This is just as well, since a macrocosm that could suddenly lose mass, fly at light speed or shrink to minute volumes would be too unstable to sustain anything. Moreover, how could the entire CSF Continuum separate from itself, and who would supply the energy necessary to effect such distortions? To experience relativity, only a part of the CSF Continuum can detach itself and alter its relationship with the rest, and the smaller that part is the easier it is to become detached and experience quantum distortions. This is our first step on the way to uniting relativity and quantum theory. As for commonplace manifestations of relativity, contrary to received wisdom, the answer is a definite yes. Remember that a commonplace manifestation means a return to non-exotic conditions, so we are now talking about open systems experiencing normal intercourse with everything else. It has already been noted that larger masses tend to outlive smaller ones, yet the medical profession recognises the principle that inactivity and over-consumption lead to obesity and a reduced life span. In our cosmic examples above we saw that there could be different types of relativistic manifestations, and so it is with the commonplace examples.

To expand a life form, by feeding it beyond its metabolic capacity, places an enormous strain on every biological system it has. Mass has increased but in a way that impacts negatively. Another factor must reduce. This cannot be volume, since over-feeding leads to obesity – increased volume – and therefore all that is left is time. This is easier to understand on a micro-cellular level; overworked cells experience increased entropic states because they cannot contain the swell of information. Furthermore, any abnormal strain placed on cellular structures weakens them, making retention of original information & structure more tenuous.

Certainly the metabolic attributes of different quantum events can differ considerably, but there are few phenomena that metabolise so efficiently that they can afford to consume over and above statutory requirements, especially when lifestyle does not include some form of physical activity sufficient to prevent retention of unwanted mass. You may wonder about the person who under-eats; surely they should live longer? Again we deal with a question of metabolism; living systems especially require some input of material to regenerate, and they are quite specific about the precise chemical requirements that will optimise regeneration. Eating too much, too little or chemically incompatible foods all increase entropy, affecting volume, mass and the endurance of continuity. Health, then, is another expression of commonplace relativistic transformation, and also reveals more evidence of Quantum Weirdness (you would expect a more cognisant species to eat sensibly, but human excesses, eating disorders and predilection for unhealthy foods is classic evidence of departure from rational expectation).

Our initial deliberations on the nature of Quantum Eventuality are nearly complete, but before examining the precise natures of T, V and M a specific working model is required for eventuality, in order to explain Q and T, V and M in all their circumstantial glory.

Term	Symbol	Definition
Event State	Es	Condition of Q at a fixed point, eg; an isochron
Event Unit	Eu	Fixed measure of eventuality eg; a queventum
Event Value	Eva	Ratio Q/D (density)
Event Density	Ed	Ratio D/Q = 1/Ev = reciprocal of event value.
Event Velocity	Evy	VM /T = Ratio of Vol-mass Index to time.
Event Concentration	Ec	TM /V = Ratio of Mass-time Index to volume.
Event Fabric	Ef	TV/M = Ratio of Time-space Index to mass.
Event Spell	Q	T V M = Temporal Edifice = $EvyT^2$
Event Zone	Ez	Q per phase = Q/Pe
Phase Event	Pe	Number of Ez in Q
Valley Event	Vae	Several linked/concurrent Temporal Edifices
Event Singularity	Esi	1/Q

Event Value (Eva) = $\dfrac{Q}{D}$ = TV^2 = ratio of eventuality to mass concentration.

Event Density (Ed) = $\dfrac{D}{Q}$ = $\dfrac{1}{Eva}$ = $\dfrac{1}{TV^2}$

Event Velocity (Evy) = $\dfrac{VM}{T}$ (gmcm^3/s) = SvyM = $\dfrac{DV^2}{T}$ = $\dfrac{Q}{T^2}$ = MvyV

Event Concentration (Ec) = $\dfrac{TM}{V}$ (gms/cm^3) = $\dfrac{1}{Sc}$ = $\dfrac{Q}{V^2}$ = $\dfrac{MvyT^2}{V}$ = DT

Event Fabric (Ef) = $\dfrac{TV}{M}$ (cm^{3}s/gm) = $\dfrac{T}{D}$ = Tva = $\dfrac{1}{Td}$

Event Singularity (Esi) = $\dfrac{1}{Q}$

The Specific Interpretation of Time

The definition of time is the endurance of continuity of mass in space. It is an attempt, albeit a crude one, to provide a regulated measurement of the metabolic velocity of temporal edifices. By nature it defines entropy devaluation, entropy osmosis and inverse entropy. The more rapidly a temporal edifice metabolises, the shorter its perceived time is and the greater its percentage of alteration per time unit. The slower a temporal edifice undergoes metabolic processes, the longer its life span, particularly its integral continuity (duration in original form).

In a total vacuum there is no matter and therefore no change, so literally there would be no passing of time. Also, if matter could be completely deactivated (prevented from metabolising even at subatomic level), then time would no longer exist for that matter and by extension the matter would cease to exist.

Our reference point in attempting to measure time must therefore be hypothetical deactivation, with a zero decay value. At this temporal state even a minute thawing of the deactivated condition becomes a noticeable corruption. It is this corruption, or deviation from the original or previous state, that we refer to as entropy. And it is marked by the first release of information from the previously deactivated matter; the first evidence of the underlying trend in all organised units to lose their organisation.

As we have noted before, entropy is a normal state of the CSF Continuum and everything in it but on the macrocosmic scale it is exactly matched by inverse entropy. As such the principle of entropy osmosis is at work moving the information and organisation of the cosmos away from some temporal edifices and to conjunction with others. Instances of inverse entropy are relatively easy to detect, partially because of their rarity which makes them conspicuous and partially because they almost always involve some form of extensive shake-up in their environment. We can see this in the amount of work and reorganisation that almost always accompanies the appearance of a new life, building or device.

Even in the realm of ideas the same fundamental truth holds; the composer who writes a monumental symphony can labour for months or even years to the detriment of his own health, in order to hone the elements of sound into the masterpiece. Great order is produced at the cost of physical comfort and relaxation, also the possibility of wresting the same patterns and organisation from sound elements in the future.

None of these examples of inverse entropy, however, amount to such a thing as temporal reversal or anti-entropy. Such a phenomenon, essentially anti-events or un-happenings, would be evidence of negative time and reverse entropy. Not only has such a circumstance never been encountered, it defies logic and all laws of physics. A lump of metal and some crude oil will never, on their own, gradually metamorphose into a car. And we should not confuse any form of maintenance with time reversal; introducing new material to a temporal edifice such as an animal, car or building may considerably extend its life span but it is achieved by premature corruption of other organised systems. Furthermore maintenance might hold back inevitable failure of a system but it is postponement, not immortality.

Our model of time, then, is significant in understanding that all temporal edifices are transitory, the CSF Continuum is eternal and that the endurance of continuity is therefore fundamentally different depending on which scale it is interpreted. On the macrocosmic scale there is only one eternal unit of time, which has never begun and will never end, because infinity has no start or finish. On the photo-cosmic scale, life and perception of it, is very different; an infinite number of temporal units segment experience into an unfolding narrative of the transitory and the temporary. Locally speaking, nothing lasts forever and it is only a matter of time before even the grandest and greatest eventualities must perish.

In quantifying our perception of time we need to expand our eventuality model;

Term	Symbol	Definition
Isochron	Ic	Fixed temporal division in phases of the cosmos
Isoflux	If	Fixed temporal division in phases of Q
Temporal Unit	Tu	Measure of event duration, eg; a second
Temporal Value	Tva	Ratio T/D = ratio of time to mass concentration.
Temporal Density	Td	Ratio D/T = 1/Tva
Temporal Velocity	Tvy	T/VM = ratio of time to Vol-mass Index.
Time-space Index	TV	$\sqrt{Q/Td}$ = MsiQ
Temporal Concentration	Tc	T/V = ratio of time to volume.
Temporal Fabric	Tf	T/M = ratio of time to mass.
Time Spell	T	$\sqrt{Q/Evy}$ = VM /Evy
Temporal Zone	Tz	T per phase
Phase Time	Pt	Number of Tz in T
Valley Time	Vta	T for linked/concurrent Temporal Edifices
Time Singularity	Tsi	1/T

Temporal Value (Tva) = $\dfrac{T}{D}$ = $\dfrac{SvyT^2}{M}$ = $\dfrac{TV}{M}$ = Ef = $\dfrac{1}{Td}$ = $\dfrac{Q}{M^2}$ = TfV = $\dfrac{Tf}{Md}$

Temporal Density (Td) = $\dfrac{D}{T}$ = $\dfrac{M}{TV}$ = $\dfrac{M^2}{Q}$ = $\dfrac{1}{Tva}$ = $\dfrac{1}{TfV}$ = $\dfrac{M}{SvyT^2}$ = $\dfrac{Md}{Tf}$

Temporal Velocity (Tvy) = $\dfrac{T}{VM}$ = $\dfrac{T}{DV^2}$ = $\dfrac{T^2}{Q}$ = $\dfrac{1}{MSvy}$ = $\dfrac{1}{VMvy}$ = $\dfrac{1}{Evy}$

Temporal Concentration (Tc) = $\dfrac{T}{V}$ = $\dfrac{1}{Svy}$ = $\dfrac{MT^2}{Q}$ = $\dfrac{Q}{MV^2}$ = $\dfrac{M}{Evy}$ = $\dfrac{Md}{Tsi}$ = $\dfrac{TMvy}{Evy}$ =

$\dfrac{Vsi}{Tsi}$

Temporal Fabric (Tf) $= \dfrac{T}{M} = \dfrac{1}{Mvy} = \dfrac{Tva}{V} = \dfrac{1}{DSvy} = \dfrac{VT^2}{Q} = \dfrac{Q}{VM^2} = \dfrac{Tc}{D} = \dfrac{Md}{Td}$

Time Singularity (Tsi) $= \dfrac{1}{T} = \dfrac{Svy}{V} = \dfrac{VM}{Q} = \dfrac{1}{TfM} = \dfrac{Md}{Tc} = \dfrac{Mvy}{M} = \sqrt{\dfrac{Evy}{Q}} = \dfrac{Evy}{VM}$

The Specific Interpretation of Space

The containment of mass through time, volume is a measure of the location and storage capacity of quantum events. A location is determined from vertical, horizontal and diagonal co-ordinates. Storage capacity is determined by dimensional factors – length x width x depth.

Term	Symbol	Definition
Vol State	Vs	Fixed dimensional co-ordinates
Isovol	Iv	Spatial division between areas of equal volume
Spatial Unit	Su	Measure of event capacity, eg; cubic centimetre
Spatial Value	Sva	Ratio V/D = ratio of volume to density.
Spatial Density	Sd	Ratio D/V = ratio of density to volume.
Spatial Velocity	Svy	V/T = ratio of volume to time.
Vol-mass Index	VM	$\sqrt{Q/Tvy}$ = TsiQ
Spatial Concentration	Sc	V/TM = ratio of volume to Mass-time Index
Spatial Fabric	Sf	V/M = ratio of volume to mass.
Volume Spell	V	$MvyTva = \sqrt{Q/Ec}$ = Sf/Msi = 1/TcTsi = 1/TfTd
Spatial Zone	Sz	V per phase
Phase Volume	Pv	Number of Sz in V
Spatial Valley	Vas	V for linked/concurrent Temporal Edifices
Spatial Singularity	Vsi	1/V = Md

Spatial Value (Sva) = $\dfrac{V}{D} = \dfrac{V^2}{M}$

Spatial Density (Sd) = $\dfrac{D}{V} = \dfrac{1}{Sva} = \dfrac{M}{V^2}$

Spatial Velocity (Svy) = $\dfrac{V}{T}$ (cm^3/s) = $\dfrac{Q}{MT^2} = \dfrac{MV^2}{Q} = \dfrac{Evy}{M} = \dfrac{Evy}{TMvy} = \dfrac{Tsi}{Vsi}$

Spatial Concentration (Sc) = $\dfrac{V}{TM} = \dfrac{V^2}{DEva} = \dfrac{V}{MvyT^2}$

Spatial Fabric (Sf) = $\dfrac{V}{M} = \dfrac{1}{D} = \dfrac{TV^2}{Q} = \dfrac{Q}{TM^2} = \dfrac{Msi}{Vsi} = \dfrac{1}{Mc} = \dfrac{V^2}{EvyT} = \dfrac{Tva}{T} = \dfrac{1}{MdM}$

Spatial Singularity (Vsi) = $\dfrac{1}{V}$ = Md

The Specific Interpretation of Mass

The physical realisation of time and space, mass is a measurement of the presence and quantity of material participating in quantum events.

Term	Symbol	Definition
Isomass	Im	Balance between Q of equal mass
Mass Unit	Mu	Fixed measure of substance, eg; gram
Mass Value	Mva	Ratio M/D = ratio of mass to density
Mass Density	Md	Ratio D/M
Mass Velocity	Mvy	M/T
Mass-time Index	TM	$\sqrt{Q/Sc}$ = VsiQ
Mass Concentration	Mc	M /V = Density (D) = $EvyT/V^2$
Mass Fabric	Mf	M /TV = ratio of mass to Time-space Index.
Mass Spell	M	TMvy = Evy/Svy = $\sqrt{Q/Tva}$ = 1/SfVsi
Mass Zone	Mz	M per phase
Phase Mass	Pm	Number of Mz in M
Mass Valley	Vam	M for linked/concurrent Temporal Edifices
Mass Singularity	Msi	1/M

Mass Value (Mva) = $\dfrac{M}{D}$ = V

Mass Density (Md) = $\dfrac{D}{M}$ = $\dfrac{1}{V}$ = $\dfrac{1}{Mva}$ = $\dfrac{1}{MvyTva}$ = $\dfrac{Msi}{Sf}$ = $\dfrac{1}{TSvy}$ = $\sqrt{\dfrac{Ec}{Q}}$ = $\dfrac{Mvy}{Evy}$ =

Vsi = TcTsi

Mass Velocity (Mvy) = $\dfrac{M}{T}$ = $\dfrac{Q}{VT^2}$ = $\dfrac{VM^2}{Q}$ = $\dfrac{D}{Tc}$ = $\dfrac{V}{Tva}$ = $\dfrac{Td}{Md}$

Mass Concentration (Mc) = $\dfrac{M}{V}$ = D = $\dfrac{EvyT}{V^2}$ = $\dfrac{Q}{TV^2}$ = $\dfrac{TM^2}{Q}$ = $\dfrac{Md}{Msi}$ = $\dfrac{Tc}{Tf}$ = $\dfrac{T}{Tva}$

Mass Fabric (Mf) = $\dfrac{M}{TV}$ = $\dfrac{D}{T}$ = Td = $\dfrac{1}{Tva}$

Mass Singularity (Msi) = $\dfrac{1}{M}$ = $\dfrac{1}{TMvy}$ = $\dfrac{Svy}{Evy}$ = $\sqrt{\dfrac{Tva}{Q}}$ = SfVsi

In Summary

The TVM model has shown that the event content of any system (quantum event or temporal edifice) is the product of multiplying time by volume by mass. The consequent eventuality equation represents the system's existence measured in queventa, where 1 queventum (Q) = $1gmcm^3s$, and furthermore we have seen how the factors of this equation are transmutations of one another. Under relativistic transformations each of the factors can be distorted into one or both of the other factors. This reflects a fundamental relationship between the factors and clearly reveals them to be permutations of the Cosmic Single Field Continuum – the macrocosm. Q is the complete expression of this field, operating across all the dimensions (temporal, spatial and material). Since space itself has three dimensions and consciousness is also a dimension (paper two will explore this matter in greater detail) there are now six dimensions accessible to our senses and reasoning. We are well on our way to finding the ten dimensions of superstring theory. We have also arrived at a point where a précis of our TVM model would be useful.

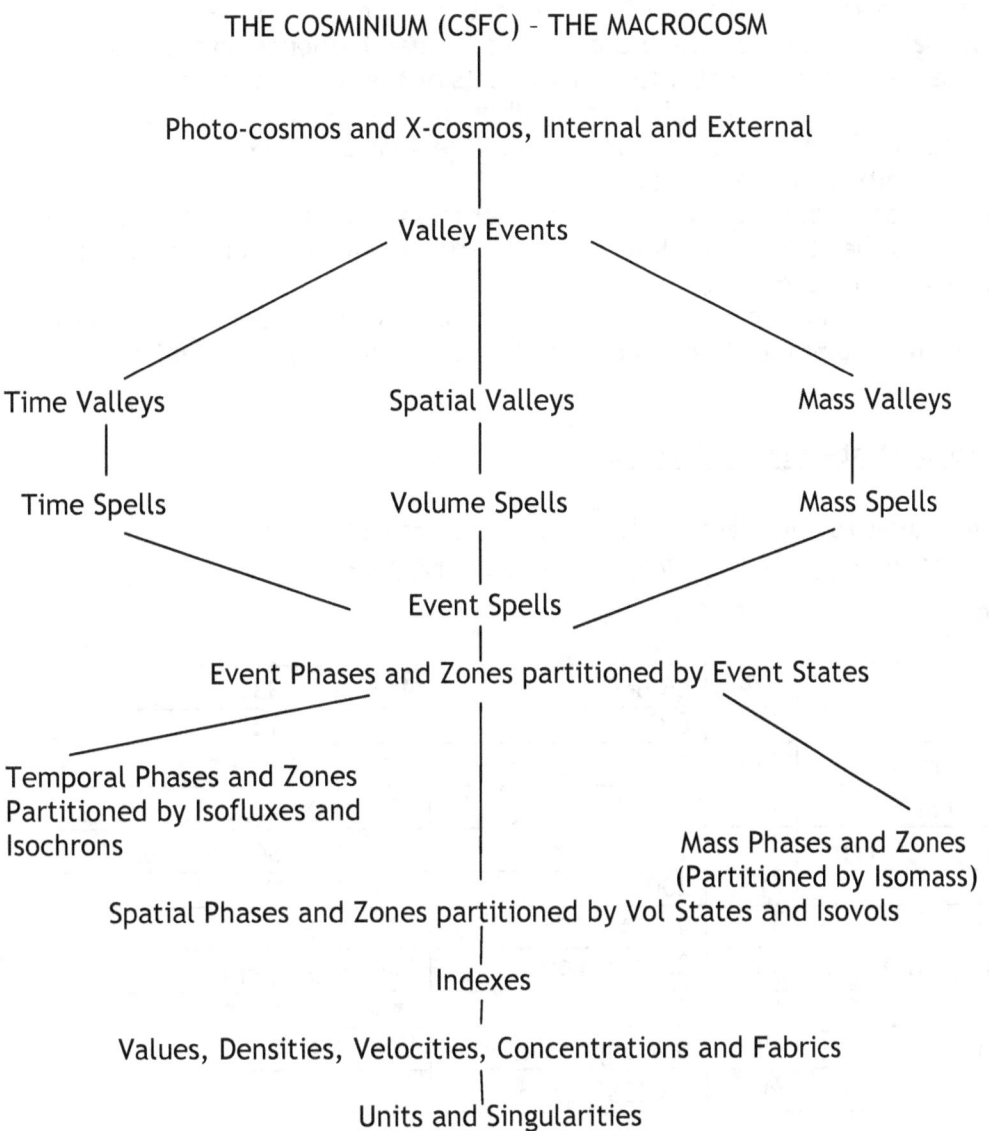

THE COSMINIUM (CSFC) – THE MACROCOSM

Photo-cosmos and X-cosmos, Internal and External

Valley Events

Time Valleys   Spatial Valleys   Mass Valleys

Time Spells   Volume Spells   Mass Spells

Event Spells

Event Phases and Zones partitioned by Event States

Temporal Phases and Zones
Partitioned by Isofluxes and
Isochrons

Mass Phases and Zones
(Partitioned by Isomass)

Spatial Phases and Zones partitioned by Vol States and Isovols

Indexes

Values, Densities, Velocities, Concentrations and Fabrics

Units and Singularities

THE TIMELAWS

1. Any collection of particles event-connected with each other, are event-independent of particles not associated with any members of the collection.
2. The expenditure or passing of time cannot exist in a vacuum, therefore neither can space nor mass. Eventuality can, however, replace vacuum.
3. Some arrangements of eventuality are more efficient at managing their payments to entropy.
4. Each quantum event is unique, a specific temporal edifice. It may be theoretically possible to duplicate it by an exact regrouping of elementary particles, but it is not possible to repeat the actual quantum event itself.
5. Particles can only follow an existence in one temporal edifice at a time. When one edifice ceases to exist its constituent particles must transfer to other temporal edifices.
6. The creation of each temporal edifice depends upon previous interactions between or within earlier edifices. Whilst the entropic demise of each temporal edifice depends entirely on its internal reactions and external interactions, it has no influence on its own origin or that of the edifices with which it interacts. It can only exert immediate influence on the present and cascaded influence on the future.
7. A temporal edifice cannot exist if the constituents of the temporal edifices that made it do not exist. Thus the existence of all quantum events is dependent upon the existence of all other quantum events. The effect cascades exponentially, proving the infinity of the macrocosm.
8. Mass distorts space-time because it affects other mass contained in that space-time.
9. Consciousness is the ability of an Internal to determine and comprehend its own interactions with the External.
10. Intelligence is the ability of an Internal to rationalise its own interactions with the External, in order to eradicate non-beneficial effects promulgating in either direction.

An Example to Illustrate more Relationships

We shall assign values to our imaginary factors; $T = 40$ seconds, $V = 50$ cm^3
$M = 30$ gm, $Q = 60000$ gmcm^{3}s, $TV = 2000$ cm^{3}s, $TM = 1200$ gms,
$VM = 1500$ gmcm^3

Eva	100000	Tva	66.666˙	Sva	83.333˙
Ed	0.00001	Td	0.015	Sd	0.012
Evy	37.5 gmcm^3/s	Tvy	0.0266˙ s/ gmcm^3	Svy	1.25 cm^3/s
Ec	24 gms/cm^3	Tc	0.8 s/ cm^3	Sc	0.04166˙ cm^3/gms
Ef	66.666˙ cm^{3}s/gm	Tf	1.333˙ s/gm	Sf	1.666˙ cm^3/gm
Pe	200	Pt	200	Pv	200
Ez	300 gmcm^{3}s	Tz	0.2 seconds	Sz	0.25 cm^3
Esi	1.666˙ x 10^{-5} Q	Tsi	0.025 seconds	Vsi	0.02 cm^3
Mva	50	Md	0.02	Mvy	0.75 gm/s
Mc	0.6 gm/cm^3	Mf	0.015 gm/ cm^{3}s	Pm	200
Mz	0.15 gm	Msi	0.0333˙ grams		

Furthermore, the difference between Q and M /TV is $(TV)^2$, between Q and T/MV is $(MV)^2$ and between Q and V/TM is $(TM)^2$. It is hardly surprising that there are so many cross connections between the three factors of eventuality. It is possible to view the factors as sides of a cube (a x b x c) and thus the geometry and trigonometry pertaining to three-dimensional objects must also apply to TVM. Therefore, mathematical rules applying to two-dimensional objects must also apply to products of any two factors (TV, TM, and VM). These cross connections exist;

Root value	1st =	2nd =	3rd =	4th =	5th =	6th =	7th =
T	EfMc	TvaMc	TcMva	EcSf			
T^2	EcEf	EcTva					
V	EvyTf	EfMvy	TvaMvy	SvaMc	Mva		
V^2	EvaTsi	EvyEf	EvyTva	Mva^2			
M	EvyTc	EcSvy	MvaMc				
M^2	EvyEc						
Q	EvaMc						
TV	EvaVsi	EvaMd	EcSva				
TM	EvaSd	EcMva					
VM	EvaTd	EvaMf					
Tc	EcMsi	EfSd	TvaSd	TfMc			
Tf	EfVsi	EfMd	TvaVsi	TvaMd	TcSf	TvyMva	
Tva	TcSva	Ef	TfMva				
Evy	MvaMvy						
Tsi	SvyVsi	SvyMd	ScMc	SfMf	MvyMsi	TdSf	
Vsi	TdTf	Md	TvyMvy	TcTsi	TfMf	SdSf	McMsi
Msi	SfVsi	SfMd	TvySvy	TcSc	TfTsi		
Td	SdSvy	Mf	MdMvy	TsiMc	MvyVsi		
Sd	MdMc	TcTd	TcMf	McVsi			
Sf	EvaEsi	EfTsi	TvaTsi	TfSvy	SvaVsi	SvaMd	MvaMsi
Svy	EvyMsi	TdSva	MvaTsi	SvaMf	SfMvy		
Sva	EfSvy	TvaSvy	SfMva				
Mvy	EvyVsi	EvyMd	TdMva	SvyMc	MvaMf		
Mc	TcMvy	EcTsi	SdMva				

The following algebra sets out general rules to cover every situation.

$$\sqrt{ab} \div \sqrt{a/b} = b$$

$$ab \div a/b = b^2 \text{ and } a(b^2) \div a/b = b^3$$

$$a^y \div y\sqrt{a} = (y\sqrt{a})^{xyy-1} \text{ for } \sqrt[y]{a} = y, a = y^y$$

$$n\sqrt{ab} \div n\sqrt{a/b} = n\sqrt{b^2} \text{ and } \sqrt[q]{b} \times b = (\sqrt[q]{b})^{a+1}$$

$$a^y b \div b^z a = a^{y-1} \div b^{z-1}$$

$$a^y / b^y = a^{y+1} b / b^{y+1} a$$

$$a^y = a^{y-2}(a-1)^2 + a^{y-2}[2(a-1)+1]$$

$$a^2 - b^2 = a + b, \text{ where } b = a - 1$$

$$a^3 - (a-1)^3 = 2b^2 + a^2 + b, \text{ where } b = a - 1$$

General structure of primary equations for values of single factors;

$(a + 1)^y = a^y + y(a^{y-1}) + x(a^{y-2}) + z(a^{y-3}) + n(a^{y-4})$ .... $+ ya + 1$

These form a Pascal's triangle;

$$(a + 1)^0 = a^0$$
$$(a + 1)^1 = a^1 + 1$$
$$(a + 1)^2 = a^2 + 2a + 1$$
$$(a + 1)^3 = a^3 + 3(a^2) + 3a + 1$$
$$(a + 1)^4 = a^4 + 4(a^3) + 6(a^2) + 4a + 1$$
$$(a + 1)^5 = a^5 + 5(a^4) + 10(a^3) + 10(a^2) + 5a + 1$$

The first ten steps in the triangle (without algebraic terms) are;

```
                1
             1     1
          1     2     1
       1     3     3     1
    1     4     6     4     1
 1     5    10    10     5     1
1   6   15   20   15    6    1
1   7   21   35   35   21   7   1
1  8  28  56  70  56  28  8  1
1  9  36  84  126  126  84  36  9  1
```

The same principles can be extended for the base unit $(a + n)$, where $b = n - 1$;

$$(a + n)^0 = a^0$$
$$(a + n)^1 = a^1 + n$$
$$(a + n)^2 = (a + b)^2 + 2(a + b) + 1$$
$$(a + n)^3 = (a + b)^3 + 3(a + b)^2 + 3(a + b) + 1$$
$$(a + n)^4 = (a + b)^4 + 4(a + b)^3 + 6(a + b)^2 + 4(a + b) + 1$$
$$(a + n)^5 = (a + b)^5 + 5(a + b)^4 + 10(a + b)^3 + 10(a + b)^2 + 5(a + b) + 1$$

And also the base unit $(a - n)$;

$$(a - n)^0 = a^0$$
$$(a - n)^1 = a^1 - n$$
$$(a - n)^2 = (a - b)^2 - 2(a - b) + 1$$
$$(a - n)^3 = (a - b)^3 - 3(a - b)^2 + 3(a - b) - 1$$
$$(a - n)^4 = (a - b)^4 - 4(a - b)^3 + 6(a - b)^2 - 4(a - b) + 1$$
$$(a - n)^5 = (a - b)^5 - 5(a - b)^4 + 10(a - b)^3 - 10(a - b)^2 + 5(a - b) - 1$$

After Pythagoras;

$(3n)^2 + (4n)^2 = (5n)^2$

$a^2 + b^2 = c$, then $(a + b)^2 = 2c - 1$

$a^2 + b^2 = c^2$, then $(a + b)^2 = 2c^2 - 1$

And generally: $an^2 + bn^2 = cn^2 = \{n(a + b - 2)\}^2$

We can attempt solution of Fermat's Last Theorem by extending the dimension;

$a^3 + b^3 + c^3 = d^3 = \{n(a + b + c - 6)\}^3$

$a^3 + b^3 = d^3 - c^3$

$b^3 + c^3 = d^3 - a^3$

$a^3 + c^3 = d^3 - b^3$

and; $a^3 \times b^3 \times c^3 = e^3 = (abc)^3$

Generally;

$a(n^3) + a + 1(n^3) + a + 2(n^3) = 3a + 3(n^3)$, so $3(n^3) + 4(n^3) + 5(n^3) = 12(n^3)$

$(an)^3 + (\{a+1\}n)^3 + (\{a+2\}n)^3 = (\{2a\}n)^3$, so $(3n)^3 + (4n)^3 + (5n)^3 = (6n)^3$

Say n = 7, a = 1 then $1(n^3) + 2(n^3) + 3(n^3) = 6(n^3)$ then 343 + 686 + 1029 = 2058

Say n = 4, a = 3 then 1728 + 4096 + 8000 = 13824

One length of our imaginary triangle is a composite number $(x^3 + y^3)$, but the solution is valid.

And the solution is reminiscent of our universal equations on page 18;

C = P + X = I + E

X = C - P + E - (P - 1)

I = C - E = (P + X) - E

Moreover these relationships are reflected in the Pascal's triangles we constructed. Interconnection is a feature of the macrocosm that cannot be ignored, and which we will be exploring more deeply in the following sections of this theory. To conclude this section I am going to briefly return to the subject of measuring eventuality. Here are some useful values to begin with;

1 Earth year = 365.256 sidereal days = 31,558,118.4 seconds

1 Lunar Month = 2,419,200 seconds

1 Week = 604,800 seconds

1 Day = 86,400 seconds

1 Hour = 3,600 seconds

1 Second = 10^{18} atto-seconds

1 cm^3 = $10^{24} Å^3$ (Ångstroms)

1 m^3 = $10^6 cm^3$

1 km^3 = $10^9 m^3$ = $10^{15} cm^3$

1 Light Year^3 = 8.4672477 x $10^{53} cm^3$

1 Parsec^3 = 2.938 x $10^{55} cm^3$ = 34.693821 Light Years^3

1 Tonne = 1000Kg = 1 Megagram

1 Teragram = 10^{12} gm = 10^9 Kg = 10^6 Tonnes

1 gm = 10^6 micrograms = 10^{18} attograms

1 electron mass = 9.10956 x 10^{-28} gm = 9.10956 x 10^{-10} attograms

1 proton mass = 16726 x 10^{-28} gm

I have spoken a great deal about Quantum Eventuality and in particular the CSFC or CSF Continuum. Now it is a good time to indulge one of the great passions of intelligence; pure reasoning. Why not calculate an estimated total eventuality for the photo-cosmos?

According to current astronomical thinking;

Age of Photo-cosmos;
18 billion years is $18 \times 10^9 \times 31{,}558{,}118.4$ seconds = $5.680461312 \times 10^{17}$ seconds

Photo-cosmos radius = 18×10^9 Light Years, therefore photo-cosmos volume =
$\frac{4}{3}\pi r^3 = 1.333^{\bullet} \times 3.141592653589 \times 5.832 \times 10^{30}$, therefore;
Photo-cosmos volume = $2.442902447 \times 10^{31}$ Light Years^3 = $2.068466012 \times 10^{85}$ cm^3

Density of photo-cosmos = approx 1 Hydrogen atom per 10,000 litres or 10^7 cm^3
Each Hydrogen atom consists of 1 electron and 1 proton,
9.10956×10^{-28} gm + 16726×10^{-28} gm = $16735.1096 \times 10^{-28}$ gm
Total number of Hydrogen masses = $2.068466012 \times 10^{85}$ cm^3 / 10^7 cm^3
= $2.068466012 \times 10^{78}$ Hydrogen masses = total mass of $3.4616005414 \times 10^{54}$ gm
Therefore photo-cosmic eventuality is;
$5.680461312 \times 10^{17}$ s x $2.068466012 \times 10^{85}$ cm^3 x $3.4616005414 \times 10^{54}$ gm
= $4.0673256509 \times 10^{157}$ gmcm^{3}s

According to my own calculations;

Age of Photo-cosmos;
5.76×10^{10} years = $1.81774762 \times 10^{18}$ seconds

Photo-cosmos radius = 28.8×10^9 Light Years, therefore photo-cosmos volume =
$\frac{4}{3}\pi r^3 = 1.333^{\bullet} \times 3.141592653589 \times 2.3887872 \times 10^{31}$, therefore
Photo-cosmos volume = $1.000612592 \times 10^{32}$ Light Years^3 = $8.472434668 \times 10^{85}$ cm^3

Total Hydrogen masses = $8.472434668 \times 10^{85}$ cm^3 / 10^7 cm^3
= $8.472434668 \times 10^{78}$ Hydrogen masses = total mass of $1.417871227 \times 10^{55}$ gm
Therefore photo-cosmic eventuality is;
$1.81774762 \times 10^{18}$ s x $8.472434668 \times 10^{85}$ cm^3 x $1.417871227 \times 10^{55}$ gm
= $2.1836277397 \times 10^{159}$ gmcm^{3}s

This is a serious quantity of eventuality, which puts us firmly in our cosmic place. Even if we were to suppose that there are a billion planets in the photo-cosmos, that possess similar creatures to ourselves, in similar numbers and of similar mass and volume, the photo-cosmos engulfs us all.
10^9 planets x 7×10^9 creatures, each living for 100 years, occupying 1.5 m^3 and weighing 80 kg;
10^9 planets x 7×10^9 creatures
x 100 x 31558118.4 seconds
x 1.5×10^6 cm^3
x 80 x 1000 gm = $2.650881946 \times 10^{39}$ gmcm^{3}s

By dividing this into the total eventuality of the photo-cosmos we obtain a ratio of
1 : 8.2373632 x 10^{119} for ourselves and our hypothetical kindred creatures, to the rest
of everything. All life is dwarfed by the reality in which it exists. And remember that in
all likelihood the photo-cosmos is greater than my estimate, and is itself dwarfed by the
macrocosm to an exponential and infinite power. In fact it is an exponential relationship
that is reminiscent of our Pascal's triangles; each row produces a bigger 'next row' with
ever increasing values, so that no matter how much of the triangle we uncover there is
always an infinitely greater amount that is hidden. Our limitations in space and time
make it impossible for us to write out anything more than the most meagre portion of
the triangle. As we shall see in the next section, the exponential foundations of
existence not only resemble Pascal's triangles; they conceal the most profound
conclusions about the true nature of reality.

As a fully participating unit of universal consciousness you want to know more, so read
on and prepare to discover even deeper revelations about how reality functions.

Section B: Aspects of Cosmic Equilibrium

Quantum Weirdness

Conventional scientific thought tells us how some particles are weightless; light bearing photons, some neutrinos and a few other particles. However, as predicted by Einstein's Relativity Theory, light and other 'weightless' forms of particle-wave energy follow a curved path around the gravitational mass of planets, stars and galaxies. As weightless as these objects may appear to our scientific instruments we know that anything affected by gravitational mass must have mass itself. Furthermore, we also know from our Quantum Eventuality model that anything that has no mass does not exist, since zero or negative values for the three factors of TVM will automatically negate eventuality itself.

Photons may have virtually no mass or volume but, however slight, they must have some in order to exist, since eventuality does not support self-contained zero values. As well as reacting to significant gravitational fields, light is blocked by opaque objects, can be slowed down in air and water, is split into components by lenses and prisms and interferes with matter. The same is true of all 'weightless' particles. Interaction and inertia are clear evidence of mass.

The taboo on zero or negative values for TVM prevents the creation of a paradoxical cosmos that would collapse as soon as it came into being. The first general principles of the CSF Continuum have already been established in section A, but reality is never as simple as a few general principles. This is what makes it an exciting and interesting place and points to the phenomenon of Quantum Weirdness – the tendency for departure from rational expectation. We have hinted at this universal characteristic already, now it is time to unveil it more dramatically.

The TVM model tells us how time moves only forward, leaving effects, memories and impressions behind it. Time and individuality are non-reversible; cause always precedes effect. We know from deduction that time travel is not possible, so it is not difficult to realise that something leaving time 'j' would suddenly acquire total entropy in that time. To the object itself entropy would appear to have ceased for the entire universe, yet suddenly at the destination time 'k' it would start again and the object's sudden materialisation would mean not only a considerable jolt of inverse entropy, but its existence alongside its own contemporary configuration. We cannot remove a quantity of matter from the macrocosm at one point and suddenly re-introduce it alongside itself at another. You might feel that this rigidity precludes the development of Quantum Weirdness, but in fact it demands it!

If Time Travel was permitted you would find yourself in a very strange, super-predetermined universe, its future a fixed and visit-able place. And if the future is fixed so are the present and past. But a history that is fixed cannot be changed and so visiting it (which necessitates changing its composition) would be impossible. This paradox means that the macrocosm is not predetermined, and so it permits the activities of Quantum Weirdness. Furthermore, there can only be one macrocosm: no parallel universes or dimensions. Why should Quantum Weirdness – the departure from rational expectation – uphold such order? Because it gives the single macrocosm and this dimension all the 'fuzziness' they need to behave precisely the way they do. It even dictates that beyond our photo-cosmos there exist other similar structures: part of an organic infinity but not separate or parallel entities at all.

Perhaps the most shocking conclusion of Quantum Weirdness is not the prohibition of time travel, the absence of mysterious alternate realities or the existence of a macrocosm pulsating with countless photo-cosmoses, but a simple redefinition of Black Hole theory. Whilst I happily accept that massive stellar objects can collapse to super-dense states that could produce a gravitational field so strong even light cannot escape, Quantum Weirdness defines not only the existence of departure from expectation, but its limitations. Even a black hole must occupy some space and have less than infinite density. Quantum Weirdness is not anarchy in the macrocosm; it is the summit and definition of uncertainty and the scope of manifestation.

Quantum Weirdness delivers both flexibility and order, and the flexibility necessary for order. In doing so it prevents the total rigidity of super predetermination that would actually destroy cosmic reason and meaning, and cause the elemental factors of infinity to freeze or coagulate until they were so brittle they could not interact with each other. Quantum Weirdness is the lubrication, the fluid medium that brings smooth interface between all parts of the whole. It provides the mechanism for osmotic entropy but, above all else, it is the very heart, soul and fuel of equilibrium. It enables the CSFC to breathe and evolve dynamically, to function at all levels and ultimately to exist. It is the exact reason why there can never be zero or negative values for TVM. And we have seen the practical implications of QW referred to in the specific interpretations of section A;

Time requires;
1. Freedom to construct all mass
2. Freedom to support all space
3. Freedom to be omnipresent

Volume requires;
4. Freedom to acquire and release mass
5. Freedom to merge boundlessly with other space
6. Freedom of temporal continuance

Mass requires;
7. Freedom of movement
8. Freedom of chemical and nuclear transformation
9. Freedom of temporal exchange

Such freedoms defy predetermination, but do require a viable structure in order to work efficiently. Furthermore, it became clear when we considered relativistic transformation, that balance was a key ingredient of existence. Having concerned ourselves with Quantum Eventuality – the main system of cosmic order - in section A, the remainder of paper one will now explore the many facets of Quantum Weirdness, which provide the freedoms.

Our universal Quantum Weirdness equation is;

$QW = Q \times rf$

Where rf = reduction factor as % deviation from Quantum Orthodoxy / 100

The Cosmic Account Mechanism

We know that infinity is all that did, does or ever will exist and that in such infinity no individual object or event (no matter how colossal) can drastically alter the total equilibrium that exists between all temporal edifices in the infinity. The most that any individual object or event can do is to create temporary disturbance in equilibrium at a local level.

In infinity, Quantum Energy Potential is limitless and existence itself is divisible by nothing; values for TVM are unrestricted and the more a photo-cosmos expands and the longer it exists, the greater its measurement of Q becomes, proving that a photo-cosmos can never be all that there is. It is vitally important that the meaning of photo-cosmos and macrocosm are properly understood. A photo-cosmos is only that part of the macrocosm that is available to the sensory perceptions of consciousness. Normally this means that it and all its contents are accessible to any creature that senses across the electromagnetic spectrum, with or without the assistance of technology. By contrast the macrocosm is infinity and as such has no beginning or end, whether temporal, spatial or material. The singularity and exclusiveness of the macrocosm are what makes other dimensions or parallel universes the idle dreams of science fiction; all things belong to the macrocosm, some within our local photo-cosmos and the rest beyond it. Anything that does not belong to the macrocosm simply does not exist.

Our first rule of Cosmic Equilibrium is therefore the Rule of Simultaneity, which is the foundation of the Cosmic Account Mechanism (CAM). It is, of course, only in a figurative way that I allude to the language of accountancy; the cosmos is a noble assembly of phenomena and therefore has no direct involvement with iniquitous man-made concepts like financial profiteering. But the structure of Cosmic Equilibrium is similar to a balance sheet in its operation and precision. It is only in this most allegorical manner that the CSFC behaves like a vast accounting mechanism, for every credit a debit and vice versa with the accounts always balancing after every transaction.

The main differences are that; the macrocosm is run by natural and incorruptible forces, whilst accountancy is subject to forces that are the creation and whim of people who are not necessarily incorruptible. And, with the macrocosm's natural laws, individual profit is dependent on successful bonding to an external environment that also profits, whereas in accountancy the welfare of external environments is deemed subordinate to individual profit.

When portions of cosmic material are transferred to new locations or submit to new conditions, Quantum Eventuality establishes a new equilibrium. CAM is present in all cosmic transactions and therefore is a vital element in the substructure of eventuality. As such it includes all values of Quantum Weirdness and Quantum Orthodoxy, requires its own structural framework and takes us deeper into the true nature of the CSF Continuum. CAM encompasses and regulates every aspect of Cosmic Equilibrium and therefore dominates all particular areas of investigation in this section.

The general equation for Quantum Orthodoxy is;

$$QO = Q - QW$$

Anatomy of the CAM

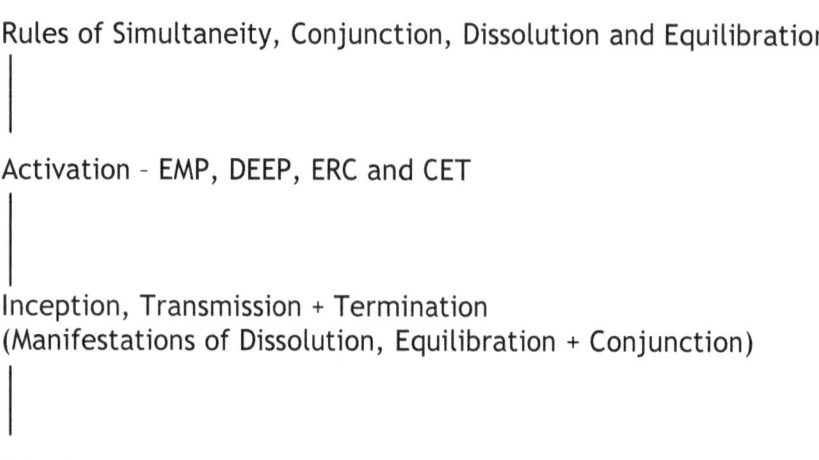

Rules of Simultaneity, Conjunction, Dissolution and Equilibration

|

Activation – EMP, DEEP, ERC and CET

|

Inception, Transmission + Termination
(Manifestations of Dissolution, Equilibration + Conjunction)

|

W + O

In order to understand the anatomy of CAM we need to consider one level at a time. Our first level can be summarised as 'The Five Axioms of CAM'.

Axiom 1: The Rule of Simultaneity and Dependency. As Q approaches zero, dependency approaches infinity, and as Q approaches infinity dependency approaches zero. Thus nothing is totally dependent on the universe, whilst the universe is totally independent. All events across the CSFC happen simultaneously.

Axiom 2: The Rule of Conjunction (Inverse Entropy). If the components of a temporal edifice are unable to undergo dissolution then they will undergo conjunction until equilibrium has been restored. Thus electrons in the third quantum shell levels of one mole of sodium atoms, mixed with one mole of chlorine atoms, cannot undergo dissolution but instead become available for conjunction with the third quantum shell levels of the chlorine atoms. The result is table salt.

Axiom 3: The Rule of Dissolution (Entropy). If the components of a temporal edifice are unable to undergo conjunction then they will undergo dissolution until equilibrium has been restored. Thus a neutron freed from an atomic nucleus becomes unstable, in which state it is unable to undergo conjunction and decays into a proton and an electron with an associated neutrino.

Axiom 4: The Rule of Equilibration. The sum of all values caused by the Rule of Conjunction equals the sum of all values caused by the Rule of Dissolution.

Axiom 5: The First Rule of Existential Proof. Phenomena that cannot undergo both conjunction and dissolution do not exist. Phenomena that undergo conjunction and dissolution must exist.
....

The second level of CAM is activation and defines specific co-ordinates in eventuality. This is the juncture at which a particular Event Reaction is born and is commonly defined by event states, isochrons, isofluxes, vol states, isovols and isomasses. We already know that TVM are powerfully connected and fundamentally interdependent factors, which, under relativistic conditions, are interchangeable. Now we are forced to face up to the inescapable conclusion that they can never be isolated factors. Whichever factor initiates an Event Reaction, all three become involved and the CAM calculations are the same. Furthermore, the 'Initiation Laws' of EMP, DEEP, ERC and CET are not factor specific; they apply regardless of whether T,V or M act as the catalyst or target of the Event Reaction.

Level three is the practical definition of Event Reactions, whilst level four contains the by now familiar concepts of Quantum Orthodoxy and Quantum Weirdness.

Forces and Fields

At present there are believed to be four cosmic force fields;

1. The Strong (binding elementary particles together in atomic nuclei)
2. The Electromagnetic (binding molecules and atoms to each other)
3. The Weak (powering radioactive decay)
4. The Gravitational (binding astronomical objects together)

Quantum Eventuality, Orthodoxy and Weirdness require the elusive Holy Grail of one all-encompassing field; the Cosmic Single Field Continuum. Scientists have struggled, ever since the early twentieth century, to find a single field theory that would resolve the apparent incompatibility between Einstein's Relativity and Max Planck's Quantum Mechanics. Theoretically such a single field theory should prove that the four forces above are really different expressions of a single binding cosmic force. Many proposals have been put forward to try and wallpaper over these cracks, most notable among the newer ideas being super-string theory.

The main conclusions of this are that there are possibly between six and ten dimensions, and the cosmos (and therefore its constituent particles) is built from these cosmic strings. Exactly what a cosmic string is like is another matter, since to the best of my knowledge we have yet to obtain direct evidence of their existence. The problems begin, however, when supporters of different types of super-string theory emerge (and they have) throwing some doubt upon the validity of the central principles. If a cosmic theory allows room for several alternate versions to vie with each other for credibility then it actually damages its own credibility. Or rather, the scientists who interpret it differently damage its credibility. A single or unified field theory should encompass all possibilities in a united vision, not spawn alternative portrayals of the universe.

Is the theory itself at fault, or is it a human propensity for conflict and supremacy that keeps us from finding the Holy Grail? With super-string theory it is still early days; there is much that is interesting and promising within it, but far too little evidence to either accept or reject it. One thing is clear, though, as it is a matter of historical precedent: as soon as anyone introduces an idea that is new to others there is always a section of the audience that is antagonistic. It does not matter how much proof is offered, how self-evident the arguments are, how elegant the concept is or how well supported by evidence it is. Someone somewhere will try to discredit it.

I prefer to keep an open mind on most issues, unless there is some logic or evidence that is beyond question. In the absence of any better method, the approach of science and logic to questions about the nature of reality is the best defence we have against ignorance. It is true that science and logic can go wrong; they are, after all, administered by human intellect that is flawed and limited at best. But the alternative is even more insecure; the unquestioning belief in the supernatural and untested. A belief in ideas that are neither supported by evidence nor rational extrapolation is at best naïve and at worst dangerous. But to pursue reality with rigorous observation and deduction is at worst a sound attempt to comprehend the universe, and at best a system that is fully self-testing, self-correcting and likeliest to reveal truth.

A million years of blind faith in magic, witchcraft, voodoo and demons will bring us no nearer to understanding reality than covering our eyes and ears will guide us safely around a dangerous, obstacle-strewn environment. We need logic, philosophy, deduction and empirical observation – the eyes and ears of thought – if we are to have a chance of navigating securely through the reality minefield.

In the absence of any greater evidence to the contrary, I propose that the methods of Quantum Eventuality and logic be applied to the question of universal forces. There is only one true Force: the Cosmic Single Field Continuum, which obeys the rules of Quantum Eventuality. For the sake of convenience, and only for that reason, we may divide it into two arenas;

Pulse Wave Interactions (PWI)
- *Including the Strong Field and*
- *The activation and increment of the Electromagnetic and Weak Fields*
- *The activation of the Gravitational Field*
- *All other nucleonic transactions.*

Universal Wave Interactions (UWI)
- *Including the transmission and increment of the Gravitational Field and*
- *The transmission of the Electromagnetic and Weak Fields*
- *All other universal transactions.*

This may seem like an oversimplification of conventional physics, or even a piece of conceptual legerdemain, but it has several immediately obvious advantages.

1. It provides a structural umbrella housing all fields within the Cosmic Single Field Continuum, and permits conceptual expansion in each arena.
2. It enables psychological acceptance of Field Unification without the need for elaborate new theories or mathematics.
3. It permits the individual constituent fields to retain their own characteristics.
4. It links the operational arenas without sacrificing the tenets of either Relativity or Quantum Mechanics.
5. It clearly demonstrates how all forces stretch across both arenas.

How the fields and forces behave is a crucial element in any understanding of the cosmos, so it should not be surprising that in a cosmos where T, V and M are permutations of the same basic function (eventuality), all forces are simply different permutations of a single force. The Strong Force is concentrated on mass but acts through space and time. The Electromagnetic Force more equally concentrates on mass and space and acts through time. The Weak Force is a substantial powerhouse of entropy and concentrates on mass and time whilst acting through space. And the Gravitational Force is the focus of time and space, acting through mass. Thus the activities and foci of these forces are distributed quite equally between PWI and UWI and all of them submit equally to CAM.

The Event Mirror Principle (EMP)

"Every Inception is balanced by a Termination of equal value and vice versa".

This is as straightforward as a balance sheet; the sides must always end up with an equal value, matching (quantum for quantum) everything that has been lost with everything gained. Our cosmic balance sheet looks like this;

Inception / Termination

-e	e
-e	e
-e	e
-e	e
- 4e	4e

The Event Mirror Principle can manifest in one of four basic types of Event Reaction;

Dispersal – Event Reaction is not reversible and Inceptions/Terminations are not interchangeable, for example, an explosion. Reaction equation;
$-e \supset \subset e$

Contrary - Event Reaction is not reversible but Inceptions/Terminations are interchangeable, for example breaking a bone. Reaction equation;
$-e \supset \leftrightarrow \subset e$

Parallel - Event Reaction is reversible but Inceptions/Terminations are not interchangeable, for example a journey in a fuel-powered vehicle. Reaction;
$-e \subset \supset e$

Reflection - Event Reaction is reversible and Inceptions/Terminations are interchangeable, for example moving an ornament on a shelf. Reaction;
$-e \leftrightarrow e$

Notice how the negative part of the reaction is always the first term in the equation. In every Event Reaction one agent must make the inception before the other agent gains the termination. This is a reflection of the universal forward direction of time and the nature of entropy. And it leads us to the expected conclusion that for every event there must have been a previous event. Specifically we are lead to the conclusion that every act of organisation follows an act of chaos, and every act of chaos must inevitably lead to an act of organisation.

Dispersal reactions are usually associated with some form of decomposition, Contrary reactions are mainly in the nature of curable damage, Parallel reactions involve spatial movement that degrades mass/energy and Reflection reactions are more likely to be spatial movements that tend to preserve mass/energy.

The Dynamic Exchange Entropy Principle (DEEP)

The greater is the magnitude of an Event Reaction, the greater are its Inceptions, Transmissions and Terminations. In CAM we learned of the first rule of Cosmic Equilibrium; the Rule of Simultaneity. Furthermore, when we first considered the Conservation Principles of the CSF Continuum, we touched on the subject of Entropic Conservation in principle No 4. It taught us that the entropy of temporal edifices is conserved through a mechanism of osmotic entropy, with isofluxes as dividing lines.

With an infinite number of simultaneous Event Reactions all achieving Cosmic Equilibrium and maintaining Entropic Conservation, it follows that the mechanism of osmotic entropy is complex and by necessity very smooth and fluid. It has to be this way in order for complex Event Reactions to take place.

Apart from individual particles, all temporal edifices will actually be a collection of smaller temporal edifices. Therefore a temporal edifice can be any quantity of eventuality from a particle (which contains only itself) up to the macrocosm (which contains an infinite number of temporal edifices). It is obvious that any temporal edifice greater than a single particle will contain an unspecified number of smaller edifices undergoing Event Reactions simultaneously. The more eventuality involved, the more complex the Event Reaction will become.

Osmotic entropy ensures that both sides of an isoflux or isochron add up to the same total before and after an Event Reaction, in order to maintain Cosmic Equilibrium. Since it is extremely unlikely that only one particle in a community of connected particles will be actively involved in Event Reactions at any one time, osmotic entropy will be active across a portion of the temporal edifice. Osmosis is a two-way flow between one side and another, with a membrane between. With the osmosis of entropy it is the isochron or isoflux that is the membrane, and entropy that is the fluid, so it follows that the transmission of entropy from one side to the other can be two-way.

It may be that in the composition of an Event Reaction there is a portion of the transmission that is positive, thus causing a partial negative result at the termination. Or it could simply be that, in undergoing the osmotic entropy, the target material that receives the Event Reaction's termination releases a small inception of its own. The two reactions are;

$$-e > xe \Rightarrow e > -xe = 0$$

$$-e \Rightarrow e + xe \Leftarrow -xe = 0$$

This, then, is the principle of Dynamic Exchange Entropy, and it drives the CAM even more dynamically than a simple Event Reaction in which entropy flows in one direction from the Inception to the Termination. Most notable of all, it helps mitigate and mediate the effects of entropy, resulting in a greater than expected life span for the temporal edifice. As well as revealing the fluidic and reversible nature of entropy, DEEP allows an incredible wealth of diverse and complex events to take place, ties in with

Quantum Weirdness and Quantum Orthodoxy and is the primary force behind the Domino or Butterfly Effect in Chaos theory.

Broadly speaking entropy is the greatest enemy of life, for life is the cosmos at its most organised and therefore it has the most to lose from entropy's undeniable influence. However, true intelligence is the greatest enemy of entropy and by true intelligence I indicate organised thought that seeks further organisation and works actively against chaos. Intelligence abhors disorder. Entropy, on the other hand, aims to achieve chaos, so that when consciousness seeks to promote entropy it is being profoundly stupid.

The Event Relativity Constant (ERC)

The anatomy of CAM, with its four rules, the activation modes, ITT and QW/QO provides a rich framework with which to comprehend and analyse Cosmic Equilibrium. Of course, all these aspects are not completely separate; the model only segregates them for our convenience. The Rules of Simultaneity, Conjunction, Dissolution and Equilibration are general properties of all temporal edifices and describe their behaviour in particular situations where they react with other temporal edifices.

The activation level then sets out how these reactionary behaviours take effect and ITT is a simple breakdown of the reaction process. Finally, when the Event Reaction is complete, we can observe the proportions of Quantum Orthodoxy and Weirdness that were involved.

Throughout this paper I have aimed to concentrate the reader's awareness of infinite simultaneous events happening throughout the CSF Continuum. The concept of simultaneity illustrates the restlessness of the macrocosm and all within it. All events are following independent evolutionary paths, yet simultaneous growth, at mostly independent rates of progress. Only when two events are in close symbiosis do their evolutionary paths synchronise. It is true that a number of events remote and unconnected to one another may independently achieve a similar evolutionary path, but this is coincidental; an example of synchronicity not synchronisation.

But for the most part each temporal edifice is unique unto itself. Consider then two edifices;
Event one: T = 54, V = 89, M = 66, Q = 317196, and Pe = 6, Ez = 52866
Factors per Ez: t = 29.71734524, v = 48.97858752, m = 36.32119973

$$t = \frac{T}{\sqrt[3]{Pe}}, \quad v = \frac{V}{\sqrt[3]{Pe}} \quad \text{and} \quad m = \frac{M}{\sqrt[3]{Pe}}$$

$\sqrt[3]{Pe} = 1.817120593$

Event two: T = 102, V = 93, M = 42, Q = 398412, and Pe = 12, Ez = 33201
Units per Ez: t = 44.5526037, v = 40.62149161, m = 18.34518976

$\sqrt[3]{Pe} = 2.289428485$

To begin with let us look at one edifice only. Event one has been divided into six bits of 52866 queventa, each containing 36.32119973 gm occupying 48.97858752 cm^3 and lasting 29.71734524 seconds. This presents us immediately with an apparent paradox.

Six units of 36.32119973 gm total 217.9271984 gm, 3.301927248 times as much mass as our original edifice. At the same time six units of 48.97858752 cm^3 give a total of 293.8715251 cm^3, again 3.301927248 times as much volume as the original edifice. Do our six separate units now last 6 x 29.71734524 seconds? That would be 178.3040714 seconds, leading us to a new total for Q of 11419056, a staggering 36 times the original, or a factor of Pe^2. Obviously this cannot be the case, so what exactly does happen and how do we resolve this conundrum?

The answer is as simple as it is elegant, and comes in two parts. Firstly, the Rule of Simultaneity – axiom 1 of the CAM – dictates that across the entire macrocosm all events experience the passing of time together. So, although we have divided our event into 6 portions, there is only one time value for all 6 portions; 29.71734524 seconds – a reduction by a factor of 1.817120593 or $\sqrt[3]{Pe}$. Secondly, we must remember that the elemental factors TVM operate together. Only time has a slightly aloof and elevated status. This means that our six portions are of Vol-Mass index, VM, and not separate portions of two factors. Six quantities of VM - 36.32119973 gm x 48.97858752 cm^3 - work out at 10673.76636 gmcm^3. We then multiply this by our new time value 29.71734524 and arrive correctly at Q = 317196.

Here then is the proof that smaller events generally have a shorter existence than larger ones. Similar results are obtained for Event two, or any other temporal edifice. Divide something up and you reduce its endurance; the more parts it is divided into the more its endurance decreases. It logically follows that if you fuse edifices together so that they behave as one, life span will increase by a factor of $\sqrt[3]{Pe}$. Let us imagine that we have ten edifices identical to Event one. Separately they will survive for 54 seconds. Together they produce a Vol-Mass index of 89 x 66 x ($\sqrt[3]{Pe}$)^2 = 27264.69281, whilst the new Q measurement is 3171960. Divide this by the Vol-Mass index and we obtain 116.3394733 seconds; an increase by a factor of 2.154434691 or $\sqrt[3]{Pe}$.

$\sqrt[3]{Pe}$ is the ERC for Conjunction or Dissolution of equal components. Compare it to the specific interpretation of eventuality in Section A, where relativistic transformations occurred. Those all involved adjustments of one or two factors in one direction and another factor in the opposite direction, with Q remaining constant. Both the situation and the mathematics were quite different, extending as far as squares and square roots. Furthermore, relativistic transformations involved single temporal edifices acting in isolation from the rest of existence whereas ERC, and the Cosmic Equilibrium rules generally, always apply to different temporal edifices or their components acting in balance with each other. The presence now of cube roots reflects the more complex nature of these interactions. Furthermore, ERC applies in several circumstances, giving rise to quite separate equations.

1. $\sqrt[3]{Pe}$ is the ERC for Conjunction or Dissolution of equal components.
2. $\sqrt[3]{Pe}$ (a) is the ERC for Conjunction or Dissolution of unequal components.
3. ERC for temporal edifices that remain separate.
4. ERC for infinity.

The first incarnation of ERC has already been dealt with, so let us look at ERC for Conjunction or Dissolution of unequal components. If we return to our examples above, let us imagine that the portions we are dividing the edifice into are different. This time we will consider Event two;

T = 102, V = 93, M = 42, Q = 398412, and Pe = 12, but Ez now is variable.

$\sqrt[3]{Pe}$ = 2.289428485 is still the same but will it apply to variable Ez?

Imagine that we divide up our Q = 398412 in the following way;

Ez1 = 1 segment of 38412 gmcm^{3}s

Ez2 = 1 segment of 36000 gmcm^{3}s

Ez3 = 10 segments of 32400 gmcm^{3}s

In this unequal split we must divide everything proportionally and calculate different values for Pe. Thus;

Pe1 = Q/Ez1 = 10.37207123
Pe2 = Q/Ez2 = 11.067
Pe3 = Q/Ez3 = 12.296666
And;

$\sqrt[3]{Pe1}$ = 2.180830087

$\sqrt[3]{Pe2}$ = 2.228486307

$\sqrt[3]{Pe3}$ = 2.308141697

Applying the rule t1 = $\dfrac{T}{\sqrt[3]{Pe1}}$, v2 = $\dfrac{V}{\sqrt[3]{Pe2}}$ and so on, the following values are obtained for the unequal segments;

Ez1 = 46.77118158 seconds x 42.64431262 cm^3 x 19.25872183 gm

Ez2 = 45.77097902 seconds x 41.73236322 cm^3 x 18.84687371 gm

Ez3 = 44.19139437 seconds x 40.29215369 cm^3 x 18.19645651 gm (x 10)

Although unequal division clearly creates complications, the ERC is still linked to cube roots of Pe or Q/Ez for each segment. Furthermore the changes remain in proportion, in the same direction, non-relativistic and linked to the total for Q of each segment. It seems that, no matter how we divide a temporal edifice, we not only split its mass and volume but significantly reduce its life span. Alternatively, by bringing separate amounts of VM into combination we give them greater endurance.

The third application of ERC relates separate temporal edifices to each other, where no conjunction or dissolution has yet occurred.

The completion point of each separate edifice is naturally the temporal value for each, so the event zone measurements are internal event reactions. Assuming both edifices begin simultaneously, when E1 has finished, the quantity of Pe2 reactions that have passed is (T1 ÷ T2) x Pe2 = 6.352941176 against 6 for Pe1, giving a completed eventuality for Event two of 6.352941176 x 33201 = 210924 at the time that Event one terminates. This is 52.9411764 % of Q for Event 2 and a Pe ratio of 1.058823529 to 1 between both edifices. This is not a comparison measure of event progress in each edifice, it directly compares the speed with which two sets of event phases occur. Let us compare this measurement with the Event Velocity for each edifice; Evy = VM ÷ T. For Event one we obtain 108.77˙ gmcm^3/s and for Event two 38.29411765 gmcm^3/s, thus Evy1/Evy2 = 2.840587129, showing the relative speeds with which events progress in the two edifices.

When Event one has expended its Vol-Mass index of 5874 gmcm^3 (108.77˙ x 6), VM consumption in Event two stands at only 2067.882353 gmcm^3, 52.9411764 % of its total Vol-Mass index (3906 gmcm^3). Apart from the obvious conclusion that Event two is more efficiently coping with entropy – it starts off with only 66% of the VM of Event one but manages to last nearly twice as long – are we any nearer to establishing an ERC for two independent temporal edifices?

To begin with we should consider the overall sum of our two edifices, but remembering that physical unity has not occurred. Total Q now stands at 715608, whilst cumulative Vol-Mass index is 9780. Because full integration has not occurred, time remains at the old levels for both events. The following ratios are thus all that can be produced;

<u>At time of Event one completion:</u>

Inferior Event Ratio (IER) = $\dfrac{(T1 \div T2)xPe2xEz2}{Q1}$ = $\dfrac{210924}{317196}$ = 0.664964249

Superior Event Ratio (SER) = $\dfrac{1}{IER}$ = 1.503840246

<u>And generally:</u>

Temporal Ratio (TR) = $\dfrac{T2}{T1}$ = IER x EVR = 1.888˙

Event Velocity Ratio (EVR) = $\dfrac{Evy1}{Evy2}$ = 2.840587129 = SER x TR

Phase Event Ratio (PER) = $\dfrac{(T1 \div T2)xPe2}{Pe1}$ = 1.058823529

$\sqrt[3]{Pe1}$ = 1.817120593

$\sqrt[3]{Pe2}$ = 2.289428485

Event Relativity Constant Ratio (ERCR) = $\dfrac{\sqrt[3]{Pe2}}{\sqrt[3]{Pe1}}$ = $\sqrt[3]{Pe2/Pe1}$ = 1.25992105

....

It is clear that the most complex situation is that involving comparisons between temporal edifices that are independent. What then of our final application for ERC – the infinite macrocosm? Surprisingly this is the simplest case of all.

ERC retains its primary definition $\sqrt[3]{Pe}$ and the value of Pe is still Q/Ez. However, it is the values of Q and Ez that have the most profound effect on this calculation. For the macrocosm we know that Q is infinite, it therefore follows that Ez is infinite so Q/Ez = 1 and our ERC formula neatly resolves as $\sqrt[3]{1}$ = 1. On the macrocosmic scale ERC is always unity, a fact that delivers the striking conclusion that as far as the cosmos is concerned all events are of equal significance. This further substantiates the ubiquity of universal law and also reinforces the realisation that comparison between temporal edifices can only be meaningful at a local level. Here I use 'local level' to mean 'similar and finite quantities'; proximity is not necessarily implied.

Cosmic Event Tension (CET)

In nature there is no conflict of wills, only the flow of eventuality itself. Reality is elastic in nature, operating in an osmotic manner across both the Pulse Wave Interactions and Universal Wave Interactions of the Cosmic Single Field Continuum. The CAM is the framework for the fine balance ruling all cosmic transactions, and within CAM we have seen how EMP, DEEP and ERC explain various aspects of event reactions. CET has a more comprehensive application and helps to tie together these different threads of CAM.

Cosmic Event Tension describes how the augmentation or diminution of a PWI or UWI resonance is always matched by an equal and opposite change in another resonance, so that CET remains at nil for the macrocosm. Tension is not an ideal description for this flowing, osmotic balance, but it does serve to convey the important principle that any deviation from zero requires resolution. CET begins with an Event Pressure Output (EPO) in an eventuality that is unable to maintain its endurance of continuity. This situation is best understood as the breach of an isoflux or isochron boundary and it causes a Cosmic Pressure Differential (CPD). It is this CPD that actually requires the resolution, more than the original EPO.

At the beginning of Section B we discussed Quantum Weirdness and Quantum Orthodoxy, looking at cosmic events on a panoramic canvas. Gradually we closed in on more specific considerations such as forces and fields, the EMP that revealed four main types of Event Reaction, and the DEEP that developed the concept of entropy as fluidic, osmotic and potentially composite. Finally ERC showed how progressions within and between event reactions obey a common mathematical relationship. With CET we return to the panoramic viewpoint with which we began, and incorporate these different aspects of Cosmic Equilibrium into a unified vision.

One of the most instantly recognisable examples of CET is the cosmic background radiation. The Mono-bloc was the ultimate EPO, causing the most gargantuan CPD imaginable – between all the matter within the Mono-bloc and the cold emptiness surrounding it. The Big Bang was the breach of the final isochron that stood between Mono-bloc and photo-cosmic evolution, and the background radiation remains as clear evidence of the CET osmosis between expanding Mono-bloc fragments and total vacuum. It shows the universe's attempt to even out the CPD to zero and produce a CET of zero.

Staying with our discussion of the Big Bang (and why shouldn't we do so in a book about cosmic eventuality?) we shall explore how all the aspects and principles of equilibrium can be illustrated in this greatest of all cosmic events.

We start with the fatal stillness at T minus 3 seconds, as outlined in the Introduction. The extreme heat and instability of the Mono-bloc stands in stark contrast to the vast expanse of empty space - the volume that will become the playing field of our photo-cosmos's galactic children. Already we can appreciate the crescendo of CET growing within the Mono-bloc, so it is clearly only a matter of time before a cataclysmic event reaction takes place. The threshold we have named the Big Bang is the ultimate isochron; a dividing line between a Mono-bloc that is far too unstable to continue, and the unfolding local universe that will become our home. The Big Bang isochron can also be seen as the terminator between ultimate order and disorder, between monumental energy potential and none.

It is perhaps problematic to try and engage the Rule of Simultaneity with our exploding Mono-bloc at T minus 0, but regardless of the smallness of that entity all that is to become the photo-cosmos is contained therein. Furthermore, with the collapse of the omnion, inequalities are already apparent, so that areas of difference exist and therefore the Rule of Simultaneity does apply. Certain sets of particles exist in combination, illustrating the Rule of Conjunction, and - since we are well aware of the impending colossal eruption - the Rule of Dissolution is also fully engaged. On all fronts the total scenario of Mono-bloc and its surrounding vacuum are either achieving or striving for equilibrium;
Extreme heat encased in extreme cold.
Gravitationally condensed mass surrounded by no mass.
Total energy set in none.
There is unlimited potential for inverse entropy, in a frame of total entropy, plus intense pressure in a huge cosmic arena devoid of pressure.

Thus the fourth axiom of CAM - the Rule of Equilibration – is also well represented, it seems. Even the fifth axiom – the Rule of Existential Proof – finds expression in a Mono-bloc embedded in icy nothingness. Rules of CAM are obviously well catered for, but what of our next level – activation?

The Mono-bloc is already its own temporal edifice and event reaction. A short-lived one admittedly, for at T minus 0 seconds it will cross a spectacular isochron and become an evolving universe, as it does so leaving behind the reign of Pulse Wave Interactions and reaching out into the realm of Universal Wave Interactions. This, then, is the main point of activation and the prime source of unbalance that is seeking to correct itself; an entire photo-cosmos of UWI matter packed into a PWI space. This is also where the CET of the Mono-bloc has been generated, and it is in the surrounding vacuum and UWI field that CET will achieve its equilibrium.

On a more local and specific level the Big Bang will see an outpouring of many millions of nuclear pieces – lumps in the very fabric of existence - as forces and energies form or lose alliances with each other and gravitational effects begin to re-assert themselves. Each 'piece' of cosmic material on the Mono-bloc side of the Big Bang isochron is therefore an Inception of the EMP, whilst each 'piece' on the expanding photo-cosmos side is a Termination. Transmission occurs as each 'bit' crosses its own isoflux. EMP is satisfied because everything that was in the Mono-bloc ends up in the photo-cosmos according to a typical pattern of Dispersal Event Reactions. The precise incident that has occurred cannot be rewound or returned to its previous condition, being of the one way -e $\supset \subset$ e variety.

DEEP must also be involved at the most extensive and impressive level, for the Big Bang is surely an Event Reaction of the greatest magnitude and complexity that our local photo-cosmos is likely to ever experience. Within this one colossal temporal edifice there are countless billions of smaller ones, at least to the extent of 1.417871227 x 10^{55} gm, which will produce an ever multiplying figure for Q as this mass continues to expand through space and extend its endurance of continuity. Osmotic entropy ensures that both sides of all isofluxes and the Big bang isochron must balance exactly. A certain amount of structure has been lost in this mighty eruption, so at least to begin with a portion of Event Reactions crossing from the PWI of the Mono-bloc to the UWI of the photo-cosmos will actually involve a transmission of entropy, a creation of disorder. This is a natural distribution of entropy, an act of balance itself, which ultimately will contribute at the most fundamental level to the Quantum Weirdness of some cosmic matter becoming aware and even intelligent.

Since Q is ever expanding following the Big Bang, we know that mostly relativistic transformations are not involved in proceedings. This means that ERC is the dominant factor at work, causing exponential leaps in both Vol-Mass Index and time according to the ERC of $\sqrt[3]{Pe}$. Eventually the current level of Q is reached - 2.1836277397 x 10^{159} gmcm^3 s. The next cosmic leap will again see an increment by the ERC equation of $\sqrt[3]{Pe}$. At each fluid stage in this unfolding cosmic drama, CET will recalibrate itself automatically. The cosmic background radiation is already sufficient evidence of this.

Only one more aspect of cosmic equilibrium requires discussion now. Quantum Energy Potentials will put a precise valuation upon the outer limits of EPO, CPD and CET.

Quantum Energy Potentials (QEP)

What is the definition of this title? We start with Einstein's most familiar relativity equation, E = mc^2 . Mass times the square of the speed of light yields the amount of energy locked in matter. Since mass is one component of eventuality, it follows that for quantum eventuality E = Qc^2 . This is our fundamental QEP and it reveals the amount of energy available from the mass contained in Q. If you remember, though, we discovered that mass, volume and time are different expressions of the same root substance; eventuality itself. All three are interchangeable under relativistic transformation. It follows that both time and volume also provide energy to the total QEP.

We shall continue this exploration by treating each factor of Q as if it is a spatial dimension in a cubic arrangement.

$\sqrt[3]{T}$ = The average time supporting Vol-Mass index in each spatial dimension.

$\sqrt[3]{V}$ = The side of a cube equal to Q, and the volume supporting Mass-Time index.

$\sqrt[3]{M}$ = The mass supported by Time-space index in each dimension.

$\sqrt[3]{Q}$ = Average eventuality supported by each spatial dimension.

$\sqrt[3]{Qc^2}$ = QEP supported by each spatial dimension.

Since the fastest possible activation of Q is at the speed of light, it follows that to measure the maximum dispersal of energy through time we must multiply Einstein's mass – energy ratio again by the speed of light;

Universal Activation Constant = c^3

The ultimate QEP of any eventuality, therefore is Qc^3, and this is the Maximum Activation Potential or MAP for short. It represents the total power that could possibly be released from a temporal edifice if even time and volume were converted into energy along with mass. At such total energy conversion any temporal edifice could interact directly and on a one-to-one basis with a comparable quantity of eventuality. MAP therefore also represents the largest portion of the universe that any temporal edifice could directly interact with.

As important as the Maximum Activation Potential is its opposite measurement; the Minimum Incumbent Dynamic (MID). To calculate this we take QEP and divide by the UAC. We then cube root the result to obtain a TVM factorisation.

$$\frac{Qc^2}{c^3} \text{ (QEP/UAC)} = \frac{Q}{c} \quad \therefore \text{ MID} = \sqrt[3]{Q/c}$$

MID represents the smallest part of a temporal edifice that will always remain active, or put another way the fraction of any eventuality that cannot become inert. To recap, then we have;

QEP = Qc^2

$\sqrt[3]{Qc^2}$ = QEP supported by each spatial dimension.

UAC = c^3

MAP = Qc^3

MID = $\sqrt[3]{Q/c}$

These are the most important Quantum Energy Potentials but there are several functions that can be derived from them.

Quantum Activity Levels

QEP per queventum = c^2

MAP per queventum = UAC = c^3

MID per queventum = $\sqrt[3]{1/c}$

Esi = Event Singularity = 1/Q = dependency of Q on each queventum

$$\frac{1}{QEP} = \frac{1}{Qc^2} = \text{ QEP Singularity = dependence of QEP on each Q}$$

MID activation levels

QEP of MID = $c^2 (\sqrt[3]{Q/c})$

MAP of MID = $c^3 (\sqrt[3]{Q/c})$

$\dfrac{1}{\sqrt[3]{Q/c}}$ = MID Singularity = dependency of MID on each of its reduced Q

$\dfrac{Q}{\sqrt[3]{Q/c}}$ = dependency of MID on each original Q

MAP activation levels

QEP of MAP = Qc^5

MID of MAP = $\sqrt[3]{Qc^2}$ = QEP supported by each spatial dimension.

$\dfrac{1}{Qc^3}$ = MAP Singularity = dependency of MAP on each queventa

$\dfrac{Q}{Qc^3} = \dfrac{1}{c^3}$ = UAC Singularity = dependency of UAC on a single queventum

$\sqrt[3]{Qc^3}$ = MAP supported by each dimension.

What exactly do activity levels tell us about the way in which the CSFC works? A great deal, in fact, because they reveal not only equilibrium relationships between the very large and the very small, but do so to an infinite regression. Furthermore, they have brought Einstein's relativity firmly into a structure that includes the sub-nuclear realm. Activity levels also provide specific lower and upper limits on entropy and reflect the universal quality of interdependence, thus tying together all non-relativistic event reactions and allowing room for organisation and the ultimate Quantum Weirdness of life.

The singularities particularly deal with dependence of one activity level on another, from Esi, which we first encountered in Section A, through MIDsi and MAPsi to UACsi. QEP, MID and MAP create a hierarchy from which follow some important conclusions.

Proof of Infinity

If you derive the MAP for any eventuality, even a single electron, and then derive the MAP for the result and continue to repeat the exercise indefinitely, there is no upper limit on the number of times MAP derivation can be carried out. Eventually you will pass the total quantum eventuality of the photo-cosmos.

This is how the sequence unfolds: (velocity of light = C = 2.997925×10^{10} cm/s)

We will assume that volume and time multiply out at unity, 1 cm^3 s.

1. 1 x electron mass = 9.10956×10^{-28} gm
2. MAP for 1 electron mass = 24544.81098 gmcm^3 s.
3. MAP of result # 2 = $6.613357243 \times 10^{35}$ gmcm^3 s
4. MAP of result # 3 = $1.781903884 \times 10^{67}$ gmcm^3 s
5. MAP of result # 4 = $4.801164273 \times 10^{98}$ gmcm^3 s

6. MAP of result # 5 = $1.293626361 \times 10^{140}$ gmcm^{3}s
7. MAP of result # 6 = $3.485548645 \times 10^{171}$ gmcm^{3}s

In only 7 steps we have moved from a single electron to something $1.596219256 \times 10^{12}$ times bigger than the photo-cosmos. This is only a factor of $1.687989519 \times 10^{19}$ off the MAP for the photo-cosmos, and an eighth step will surpass even that. Imagine the amount of eventuality you could reach in a thousand steps! Clearly there is much more to the universe than the portion accessible to our perception.

Proof of non-infinite Regression

Starting with an eventuality of any size and deriving MID calculations also creates some interesting outcomes. This time we will begin with my rough calculation for Q of the photo-cosmos.
1. Photo-cosmos = $2.1836277397 \times 10^{159}$ gmcm^{3}s
2. MID of Photo-cosmos = $8.997446638 \times 10^{49}$ gmcm^{3}s
3. MID of result # 2 = $1.442445806 \times 10^{13}$ gmcm^{3}s
4. MID of result # 3 = 7.835972692 gmcm^{3}s
5. MID of result # 4 = $6.393775503 \times 10^{-4}$ gmcm^{3}s
6. MID of result # 5 = $2.773185196 \times 10^{-5}$ gmcm^{3}s
7. MID of result # 6 = $9.743598274 \times 10^{-6}$ gmcm^{3}s

You may notice that as the results have dipped below unity and into negative powers, the results for MID are getting nearer to the previous MID. By the time the outcome is in the realm of the electron, MID is actually greater than the electron. There may be something less than the electron, the equations are telling us, but you can go no further in subdividing the fundamentals. We have arrived at the kingdom of the Pulse-Wave (see section D).

Illustration of Quantum Energy Potentials for planet Earth

Age to date = 1.5779059×10^{17} seconds; Mass = 5.977×10^{27} gm

Radius = 6371.23 km = 637123000 cm, Volume = $\frac{4}{3}\pi r^3$ = $1.083324236 \times 10^{27}$ cm^3

Eventuality to date (Q) = $1.02169864 \times 10^{72}$ gmcm^{3}s

QEP calculations for Earth

QEP = Qc^2 = $9.182572011 \times 10^{92}$

$\sqrt[3]{Qc^2}$ = QEP per dimension = $9.719742978 \times 10^{30}$

c^2 = $8.987554306 \times 10^{20}$ = QEP per Q

QEPsi = $\dfrac{1}{Qc^2}$ = $1.089019502 \times 10^{-93}$

QEP of MID = $c^2 (\sqrt[3]{Q/c})$ = $2.913906047 \times 10^{41}$

QEP of MAP = Qc^5 = 2.474153464 x 10^{124}

MID calculations for Earth

MID = $\sqrt[3]{Q/c}$ = 3.242156818 x 10^{20}

MID per Q = $\sqrt[3]{1/c}$ = 3.219040288 x 10^{-04}

MCP = $Q\sqrt[3]{Q/c}$ = 3.288889084 x 10^{68}

MIDsi = $\dfrac{1}{\sqrt[3]{Q/c}}$ = 3.084366538 x 10^{-21}

MID of MAP = $\sqrt[3]{Qc^2}$ = 9.719742978 x 10^{30}

$\dfrac{Q}{\sqrt[3]{Q/c}}$ = 3.151293097 x 10^{51} = dependency of MID on Q

MAP calculations for Earth

MAP = Qc^3 = 2.752866219 x 10^{103}

c^3 = UAC = 2.694401374 x 10^{31} = MAP per queventum

MAP of MID = $c^3 (\sqrt[3]{Q/c})$ = 8.735671785 x 10^{51}

MAPsi = $1/Qc^3$ = 3.632577541 x 10^{-104}

$\sqrt[3]{Qc^3}$ = MAP supported by each dimension = 3.01945366 x 10^{34}

General calculations for Earth

Esi = $\dfrac{1}{Q}$ = 9.787621916 x 10^{-73}

UACsi = $\dfrac{1}{c^3}$ = 3.711399533 x 10^{-32}

$\sqrt[3]{T}$ = 540373.0732

$\sqrt[3]{V}$ = 1027037150

$\sqrt[3]{M}$ = 1814795743

$\sqrt[3]{Q}$ = 1.007181187 x 10^{24}

The Specific Interpretation of Energy potentials and Activity Levels

What do the above QEP calculations tell us about Earth?

The QEP and MAP for Earth, like any other quantity of eventuality, reveal how much power could be realised if Q was turned to pure energy. This could involve not only the conversion of mass to energy but also the relativistic transformation of time and space. As with all universal phenomena, there are different ways for eventualities to mature, so QEP could just as easily reveal how much total energy the Earth would have given out over its life span and dimension, if it had been pouring out its matter as radiation, like a star. Here is another proof of the interconnectivity of T, V and M; if Earth had been irradiating itself away as total energy, not only would its matter have declined – so would its volume and remaining life span.

The figure for QEP supported by each spatial dimension $\sqrt[3]{Qc^2}$ is a guideline figure only and assumes that the Q being considered is cubic. This then is the QEP contributed by each side of the (hypothetical) cube. Although it may only be an imaginary metrication it serves to illustrate an important principle; that space is a geometric extension in three vital dimensions. Each dimensional extension supports eventuality to the extent of this calculation and thus volume is not separate from mass or time, but intimately interwoven with it. If our comfortable 3D world collapsed there would be grave consequences for eventuality.

QEP per queventum - c^2 - is another constant. Einstein has already told us all we need to know about its participation in the scheme of things.

QEPsi is an interesting derivation. The reciprocal of QEP, it informs us of the extent to which QEP is dependent on each queventum of a temporal edifice. Since QEP is larger than Q, QEPsi is inversely smaller than Esi. In a nutshell it is easier to coax energy from larger masses, and more matter means more potential release. QEPsi also returns us to the notion of greater mass, greater longevity; larger masses miss released energy (entropic decay) less then smaller masses. For a minute mass of $\dfrac{1}{c^2}$ gmcm^{3}s both its QEP and its QEPsi will be 1 – showing that it is totally dependent on its own minute eventuality.

QEP of MID is interesting because it arrives at a figure near to MID squared and suggests the immediate reactivation potential of MID.

QEP of MAP brings Earth's universal participation remarkably close to the entire photo-cosmos.

What do the above MID calculations tell us about Earth?

Assuming that the Earth's average density is consistent (and it is not – density increases nearer the core) we have a figure of 5.517277101 gm/cm^3. A rough volume for the Earth's core is 5.015389985 x 10^{24} cm^3. This yields an estimated core mass of 2.767129632 x 10^{25} gm. The age for the core is the same as the whole Earth, so measurement of Q for the Earth's core is 2.189854769 x 10^{67} gmcm^{3}s. Our MID figure of 3.242156818 x 10^{20} is therefore only 1.480535086 x 10^{-41} % of the Earth's core. And it is 3.173300513 x 10^{-50} % of total Earth Q to date. Even if Earth was flung out into deep space billions of light years from any heat source, this is the minimum portion of our planet that would remain active.

The interesting thing about calculating a MID value for any unspecified quantity of Q is that the larger Q is the smaller the proportion of it is that represents its MID value. Alternatively, the smaller Q is the greater a percentage of it is the MID value. As a stock part of the MID calculations, MID per Q is $\sqrt[3]{1/c}$ = 3.219040288 x 10^{-04}, a value that is true for all temporal edifices and can therefore be truly classified as a universal constant. As you can see this is a much more significant percentage of the original single queventum than the MID figure for Earth. Another interesting outcome is to multiply this constant by the Q calculation for Earth; the result is 3.288889084 x 10^{68}, which is remarkably close to our estimate for the Earth's core. This, then, is the Minimum Constant Potential or MCP, and represents the lower threshold dynamic of any temporal edifice, rather than the true MID. It is a useful additional measurement in understanding how eventuality behaves.

MIDsi, the MID singularity, simply measures the dependency of MID on each of its own reduced number of queventa. $\dfrac{Q}{\sqrt[3]{Q/c}}$ on the other hand measures MID dependency on each Q of the original edifice. MIDsi is greater than Esi so MID is more dependent on its own queventa than the original Q on its own. This suggests that if an edifice such as Earth was quantum frozen, its MID would only retain activity if it remained in contact with the rest. Again we are shown that larger temporal edifices can expect greater endurance of continuity, whilst separating parts of a thing from the whole is asking for a reduction in longevity.

MID of MAP is identical to QEP per dimension, but in this context reflects what would happen if original Q was ignited to MAP level, then deactivated. It is a little reminiscent of a stellar nova explosion, where a star casts off most of its mass and is then left as a mere shadow of its former self.

What do the above MAP calculations tell us about Earth?

MAP itself is the extent of possible immediate interaction that an edifice may have with the rest of the universe. For Earth we are looking at a quantity of eventuality roughly equivalent to our home galaxy, which is really not surprising. What is alarming is that this figure also represents the extent of entropic damage Earth could do if it was energy converted to the ultimate level, beyond even QEP. Whilst QEP speaks of the liberation of mass to energy, MAP speaks of Q's transformation to super energy or Pulse Waves, drawing power even from temporal and spatial definitions.

MAP of MID predictably gives us an ignition of MID, again bringing us close to our MCP calculation for the Earth's core.

UAC – the Universal Activation Constant – represents this same anarchistic transformation for each queventum.

MAPsi, or MAP singularity, is the dependence of MAP on each queventum, and as one might expect it is minuscule.

$\sqrt[3]{Qc^3}$ is MAP supported by each dimension, returning again to our hypothetical cube and the eventuality burden carried by each of its dimensions. As you can see, it is not unlike the MID of MAP.

What do the above general calculations tell us about Earth?

Esi is a basic dependency ratio for Q against each of its queventa. There are no prizes for seeing that greater values of Q create lesser dependency on each $gmcm^3$s present. The word is more impacted by the loss of the text than the text is by the loss of the word.

What is of particular interest is our UACsi figure. We have another constant; UAC singularity is identical for queventa in all temporal edifices and reveals UAC's dependency on a single queventum. At heart all queventa are identical until they undergo conjunction or dissolution. Like any other temporal edifice, at its most fundamental level Earth is made of nothing spectacular; distinctiveness is entirely the result of embellishment and juxtaposition.

The cube roots of TVM and Q are the quantities of those factors that would be provided by each dimension of V if the temporal event was cubic in shape. This is a notional calculation but it is helpful in equating quantum events of different topological shapes and/or varying distributions of elemental factors.

How do Quantum Energy Potentials advance the cause of Cosmic Equilibrium or our understanding of it? They relate the large to the small and in particular the very large to the very small. There is a clearly defined hierarchy as events climb from PWI to UWI, but underlying all things are the constants binding the CSFC together.

ERC = $\sqrt[3]{Pe}$,

Event Relativity Constant Ratio (ERCR) = $\sqrt[3]{Pe2/Pe1}$

MAP/Q = UAC = c^3 , UACsi = $1/c^3$ = 3.711399533 x 10^{-32}

QEP/Q = c^2 = 8.987554306 x 10^{20} , MID/Q = $\sqrt[3]{1/c}$ = 3.219040288 x 10^{-4}

Section C: Quantum Eventuality Architecture

Definitions and Preliminary Considerations

Before proceeding with the next phase of this investigation I should quickly recap some of the phraseology and descriptions that have guided us along our journey so far.

<u>The Macrocosm</u>

This is a term for the infinity of time, space and matter. The macrocosm has no beginning or end and no boundary. It stretches back in time further than the imagination, beyond the Big Bang into a limitless history. Its endurance of continuity has no upper limit, so there is no entropy that can ever harm it. Its reach through space is complementary so that no matter how far or how fast one could travel, or for how long – even an eternity – you would never encounter an edge or boundary. And its containment of matter is similarly without restriction. Regardless of how thinly substance may be extruded throughout the limitless volume, there is still endless mass. We have already spoken of the infinite grid of cubes; no matter the size of the cubes or the paucity of matter in each, endless volume leads to infinite matter.

This is further substantiated by relativistic transformation; it should come as no surprise that matter is infinite, since it is merely one possible expression of eventuality. Since time and volume are infinite, and matter is interchangeable with them, it follows that matter also has no end. In fact all three elemental factors of infinity are interchangeable, being different expressions of a single source – eventuality itself. At the root of all things, it is therefore eventuality itself that is infinite and TVM that are three interchangeable manifestations of that infinity. The infinite operating medium of eventuality is the Cosmic Single Field Continuum, a unified wave function of existence itself. At the nuclear level CSFC can be observed as Pulse Wave Interactions and at super nuclear levels the continuum evolves into Universal Wave Interactions.

On a scale we can readily comprehend, the Macrocosm can be hypothetically divided in two ways;

The X-cosmos + the Photo-cosmos or
The External + the Internal

For convenience we measure the infinite eventuality of the macrocosm as packets of eventuality called queventa - $gmcm^3s$, and experience them as quantum assemblies called temporal edifices. There is a certain fallacy in doing this, because the reality is that infinity cannot be divided, but our own finite nature makes it impossible for us to adequately contemplate the unbounded. Yet it is our grounding in the finite that enables us to envision the relationships of finite phenomena to each other.

A powerful tool in interpreting the deluge of information from infinity, in a finite way, is to extract and apply mathematical relationships in a finite manner from the minute to the gargantuan scale. As we have just seen in Section B, a number of mathematical constants can be derived and a complex Cosmic Accounting Mechanism calculated to express local applications of the cosmic equilibrium. By understanding events on a local platform we hope to extrapolate to the panoramic.

The X-Cosmos

That part of the Macrocosm that lies beyond the Photo-Cosmos. The X-cosmos is forever beyond the reach of perception.

The Photo-Cosmos

That part of the Macrocosm we can actually 'see' (hence photo) either with our own senses or sense-enhancing scientific apparatus. For various reasons I am peeling away gradually, it is a mistake to believe that what is visible to our limited perceptions is all that there is. And there is always the searching philosophical question "if the visible universe is everything then it has no boundary; if it has no boundary then there is something beyond the range of our perception, in which case the visible universe is not all that there is".

The External

This is that part of reality that is not the self. By extrapolation this includes the X-cosmos and all of the photo-cosmos apart from the observer (Internal). And it is a foregone conclusion that the External is slightly different for each of us. For you the External includes me, whilst for me the External includes you. It is a matter of straightforward deduction that, since the External is never exactly identical for any two observers, an absolute comprehension of reality is beyond the intellectual capacity of all observers, both individually and collectively. Moreover, since the External includes an unspecified number of conscious organisms, it also includes us, each and every Internal that there is.

We cannot fully know the External, therefore we cannot fully know ourselves. But there is one ray of hope in our situation; we react with the External, as any temporal edifice does. We exchange substance and information with what is beyond, in an exquisite and sometimes painful osmosis. Over time we literally undergo chemical and physical metamorphosis. For consciousness that is sensitive and capable of growth, there is also substantial intellectual and personality metamorphosis. This is the crux of life and, as I have said before, entropy is life's main enemy but intelligence is the curse of entropy, and thus our best ally. For consciousness to behave in a destructive way, eschewing intelligence, is mad, calamitous, stupid and mostly tragic.

The Internal

This refers to the self; an individual or observer. Each unit of consciousness is an Internal, swimming in the sea of reality. The less an Internal understands that sea the more likely it is to drown. And it follows that the more of that sea that is understood the better, particularly those parts that are less predictable - such as other Internals. It stands to reason that if two or more Internals are engaged in direct and intentional conflict, it will be a miracle if either survives. And survival is life. We are all tied together by reality, and when any part of reality is lost to the depths there is the possibility that it will pull us into the abyss too. In order to bring that loss about ourselves we need to get close to the target, thus increasing the danger. Please remember that every time you are responsible for destroying consciousness by design or negligence, as when you eat animal flesh.

Entropy

Entropy is the decay through time of all finite phenomena. Decay is defined as loss of information, energy or structural integrity – most typically all three. If left alone for a sufficient length of time a temporal edifice will experience entropy as a slow and gradual process culminating in an inability to maintain sufficient pattern and organisation to continue functioning as a whole. The precise internal organisation of an edifice may result in greater efficiency at resisting entropy, whilst larger edifices tend to hold together longer by mere strength of association, but eventually all finite things will reach a point of no return. It is also true that continual introduction of beneficial matter to an edifice will prolong its life, by replacing worn, damaged or redundant components. But this is still only a matter of postponement; the end will come to all things except the macrocosm itself.

Only infinity avoids succumbing to corrosion or decomposition, and it does so by endless reshuffling of its infinite components. As parts of reality fade away their essence is recycled into new arrangements, in an endless carousel. There may be a level of existence where one could read or interpret the history of each and every particle, and it would be a fascinating narrative if one had endless time. The question is 'how could we access this history log of matter, decipher the code and publish the results in a brief enough format to be digestible?'
……..

The resolution of our quest for this Holy Grail of universal wisdom involves finding a key, code - book and technique for condensing a gargantuan stockpile of information. The predicament can be illustrated by considering something rather less ambitious than a detailed particle history of the entire macrocosm.

Can we understand a train? Before answering, we should consider the question properly. Understanding a thing has a lot more to it than merely seeing or perceiving the thing and assuming that what is seen is what there is. In everything we encounter there is a massive hidden reality 'iceberg' that we either do not consider or are not even aware of. To really understand something we must have precise information about every single particle and event reaction it contains. Imagine the difference between seeing a sandy beach and knowing intimate details about every single grain of sand – their shapes, molecular compositions, numerical and topological relationships, energy displacements, light diffraction, ages and event histories, origins, particle configurations, fracture points, isotope compositions, boiling points, crystal structures and electrical potentials.

With our imaginary train it may help for us to be an Einstein or an IK Brunel, a master physicist or engineer. But even so we would be unlikely to produce a description anywhere close to the actual situation; it seems that comprehension at the required standard is not merely fractionally or digitally beyond our capacity, but exponentially so. A mind far more complex, intricate and potent than even the cream of human intellect may just have the processing and cognitive power to undertake our little train analysis, or even the beach test. Perhaps a creature that thinks and senses in multiple dimensions and routinely calculates up to base 3000? But even the perceptual and intuitive ability to register and process such intricate and voluminous data may not be sufficient to accurately track and analyse billions of simultaneous event reactions that are changing all the time.

The universe's greatest supercomputer might just possess the processing capacity, but would it have any understanding of the stream of information flowing through its circuits? It is indicated that the human mind at its most formidable engagement with the cosmos can only just grapple with the superficial skin of reality, much less cope with the full flesh. At only an average engagement with the cosmos we do not even scratch the skin. Either way there is a massive void in technical competency. And to a certain extent this void is not something we can easily compensate for with a simple act of faith, willpower or conventional approach. Thankfully there are some unconventional approaches, but they each have their own dangers and limitations. They are; sample analysis, logical projection and radical extrapolation. We may more comfortably and with greater familiarity know them as observation, deduction and conjecture.

In the previous section we discovered many aspects of cosmic equilibrium, concluding with Quantum Energy Potentials. In each calculation of a potential we considered the result of projecting activity to 100% efficiency. The possibility, however, of QEP functions realising 100% of their activation threshold is actually quite minor; Quantum Weirdness is ever present, along with its dark apprentice, entropy. If we allocate symbol Φ to represent any QEP function, then outcomes of less than 100% follow the equation $r = \Phi \times (p/100)$, where Φ is an increase, and
$r = \Phi \times (100/p)$, where Φ is a decrease.

r = new result
p = % activation

We know that MAP never reaches infinity, so there is no limit to how many times we can keep applying MAP. MID reduction slows with each level lower than 1 and eventually produces values close to the original figure for Q. We have reached the realm of the Pulse Wave. A number of conclusions have been obtained through MAP and MID;

1. MAP applies endlessly to infinity, supporting a macrocosm that is totally independent and all encompassing; it has no outside.
2. MID reduces to a point of no return, proving that regression is finite and that nothing is totally dependent on anything else.
3. The less Q there is the more its existence depends on outside forces, the greater its entropy value and the less information it contains. More Q means less dependence on outside forces, less entropy and more information.
4. Lower Q values carry higher entropy to information ratios and increased randomness, producing notable manifestations of the Heisenberg Uncertainty Principle and quantum weirdness. At minute levels they become extremely sensitive to any environmental conditions.
5. Larger Q values carry lower entropy to information ratios and reduced randomness, producing notable predictability curves. At enormous levels they wield considerable influence on their environment whilst being considerably resistant to external forces.
6. Lower Q values try to equilibrate by conjunction whilst larger Q values tend to equilibrate by dissolution.
7. Cosmic equilibrium is achieved through continual trade-offs between areas of low and high event tension.

"I doubt, therefore I think; I think, therefore I am", is a familiar axiom by Renée Descartes, which is amply reinforced by the principles of quantum eventuality. There is a simple proof of existence: if nothing exists then entropy is infinite and information zero. We know, however, that entropy can never be infinite whilst information can never be zero. In our CAM model this is axiom 5; **The First Rule of Existential Proof.** Phenomena that cannot undergo both conjunction and dissolution do not exist. Phenomena that undergo conjunction and dissolution must exist. By thinking about the Universe we interact with the largest body of information, thus proving our own existence to the greatest extent possible.

Our first Rule of Quantum Eventuality Architecture is an important extension of axiom 5 in the CAM; that which fails axiom 4 (The Rule of Equilibration) or axiom 1 (The Rule of Simultaneity and Dependency) does not exist.
This is The Second Rule of Existential Proof.

The second Rule of QEA is that all that exists can be defined by a quantum eventuality measurement. This may not be the only way that a particular phenomenon can be measured, but that which fails to yield a Q measurement – because it lacks one or more of the elemental factors T, V or M – is not real.
This is The Third Rule of Existential Proof.

The third Rule of QEA is that Time, Volume and Mass are all interchangeable dimensions of a single component; eventuality itself. As noted previously, we call this relativistic transformation. It mirrors the manner in which length, width and height are all interchangeable dimensions of space, as a result of rotational transformation. Anything that cannot undergo relativistic transformation does not exist, thus **this is The Fourth Rule of Existential Proof.**

The TVM hierarchy is fundamental in constructing QEA, beginning with the least important and most easily manipulated elemental factor; Mass. Mass is readily altered by simple physical interactions, but volume alterations require specific energy interactions, whilst time is only affected by relativistic forces or space-time gravitational curvature. Since T, V & M are each infinite for the Macrocosm, the ratio 1:1:1 is self-evident, firmly demonstrating that the Macrocosm is a unity. We can classify its temporal edifices as either;

DYNAMIC – causing actual positive or negative disturbance mainly to the CSFC.
PASSIVE – causing actual positive or negative disturbance mainly to itself.

It is a logical deduction that PASSIVE temporal edifices are in some way inert, and this definition includes any phenomena undergoing relativistic transformations or other relativistic reactions that temporarily disconnect them from the rest of the CSFC. It therefore follows that nothing remains inert forever, and small particles are likeliest to be PASSIVE. Nothing is an infinite PASSIVE.

Similarly we can conclude that DYNAMIC temporal edifices are not temporarily disconnected from the CSFC and are not undergoing relativistic transformations or other relativistic reactions. DYNAMIC edifices are likeliest to be large aggregates of matter. The Macrocosm is therefore an infinite DYNAMIC.

It is possible for a DYNAMIC edifice to temporarily become PASSIVE, or to possess both PASSIVE and DYNAMIC attributes, but it is not possible for an edifice to be neither PASSIVE nor DYNAMIC. **This is The Fifth Rule of Existential Proof.**

To recap, then:
The First Rule of Existential Proof.

Real temporal edifices obey conjunction and dissolution rules.

The Second Rule of Existential Proof.

Real temporal edifices obey equilibration and simultaneity / dependency rules.

The Third Rule of Existential Proof.

Real temporal edifices can be defined by quantum eventuality measurement.

The Fourth Rule of Existential Proof.

T, V & M are interchangeable dimensions of eventuality.

The Fifth Rule of Existential Proof.

Real temporal edifices are DYNAMIC and/or PASSIVE.

....

Therefore anything that cannot undergo conjunction, dissolution or equilibration, establish a level of dependence, exist simultaneously with the rest of the Macrocosm, be defined by measuring its quantum eventuality, does not possess interchangeable TVM factors or is neither DYNAMIC nor PASSIVE, simply does not exist.

Where do ideas and thoughts fit in, then? We need to make a careful distinction between reality and thought. In speaking of reality we mean to indicate whatever is definitely true. The quotation of Descartes illustrates the manner in which we must apply deductive reasoning to establish what is. We can be certain that we – the Internals – exist, and that there is something beyond us, the External. What is open to question is the precise natures of the Internal and External. Nihilists are fond of disputing any definite view of the External; but their reasoning fails in making the assumption that the External has no certainty, but that their own view does. Nihilism is self-negating and self-defeating. Dogmatic viewpoints also fail by ignoring anything beyond their own evidence island, assuming that challenging information is automatically incorrect.

Truth lies between these two extremes; on larger scales eventuality is easier to extract fairly predictable sequences from, but on a small-scale quantum weirdness interferes too much with prediction and the element of randomness prevails. Of course, there is always an element of quantum weirdness in everything. But it is only ever a portion and never a totality: the CSFC is neither dictatorship nor anarchy. In answering our question about ideas and thoughts we can be certain of only one thing; between reality and our perception and conception of it there is a gulf that is closed only by knowledge and rationality. As we will see in Paper Two, there are many stages at which the transmission of truth to consciousness can fail. We know that thoughts and ideas are real in the sense that we have them and we exist. They occupy time but do they occupy space or have mass?

Quantum Relativity Matrix

Aspect	Small scale eventuality	Large scale eventuality
Entropy	Higher	Lower
Equilibration by	Conjunction	Dissolution
Information threshold	Lower	Higher
Autonomy	Dependent/vulnerable	Independent/resilient
Influence	Local	Panoramic
Character	Quantum weirdness	Quantum orthodoxy
Determinism	Unpredictable	Predictable
TVM factor values	Negligible ≤ 1	Sizeable ≥ 1
EMP	Negative Inception	Positive Termination
Forces and Fields	Pulse Wave Interactions	Universal Wave Interactions
DEEP	Low ITT – simple reaction	High ITT – complex reaction
ERC	Simple relationship	Complex relationship
CET	Lower	Higher
QEP	MID – non infinite regression	MAP – endless to infinity
Participation	Internal	External
Cosmic division	Particles	Macrocosm
Probability	Low ratio	High ratio

Notwithstanding all of the above, are we any nearer to understanding how ideas and thoughts can exist yet apparently have neither volume nor mass? It could be argued that even such ephemeral things as these do occupy space - the connections between synapses in the brain, electrons moving between synapses; certainly an occupation of space is involved. With volume matter also becomes more real. Yet opponents to this view would say that such matter and space exist any way, regardless of whether any ideas or thoughts are passing through them, or indeed what the nature of those constructions may be.

An actual thought or concept, idea or imagination, may apparently be completely unreal after all; this would tend to agree with our definition of reality as a product of T, V & M. Are we facing another paradox? I do not believe so, because on a purely basic level the proof of space and matter being involved in the mind that thinks, are evidence enough. To add weight to this, though, I would call forth the following witnesses for the defense.

1. A thought, idea or imagination is a way of communicating information. Anything that communicates information has a finite entropic value and therefore exists.

2. Ideas themselves can only originate from a consciousness, which has physical substance and a measurable volume.

3. When an idea or thought has been expressed it leads to consequences, either within the consciousness that expressed it or - more commonly in both that consciousness and others. It is therefore something that can be reacted to.

4. Once expressed, it may be true that the thought or idea has been released from the originating mind and therefore no longer appears to be contained by any volume or matter. Where has it gone, especially if no other minds pick it up? There are two answers; either it is recorded on paper, or in a computer or on some other medium - thus it has been successfully transferred from one place to another, and can be received by other minds later. Or it has not been recorded, is forgotten by its originator and thus ceases to exist, proving our definition of what is real.

5. If T, V & M are different dimensions of a single phenomenon - eventuality - and are interchangeable with one another, then thought is simply a next dimension where V & M have undergone relativistic transformation into time and thus no longer require more than a nominal value.

These musings are not our main concern, in the Quantum Relativity Matrix, although they do participate in the overall picture and help to 'ease in' the main areas of concern for Paper Two. The central focus of Quantum Relativity Matrix is the connection between the microcosm and macrocosm, across the CSFC, in particular a probability curve relating the worlds of Quantum Mechanics and Heisenberg Uncertainties to Einstein events on the grandest of scales.

To further comprehend the interconnectedness of the small and huge scales, it is possible to extract a probability distribution diagram that shows the likelihood of different high or low value combinations of T, V & M.

Where small (S) = ≤ 1 and large (L) = > 1

Possible configurations are (in order T, V, M);

SSS outcome Q = ≤ 1
SSL outcome Q = ≤ 1 to > 1
SLS outcome Q = ≤ 1 to > 1
LSS outcome Q = ≤ 1 to > 1
LLL outcome Q = > 1 to $<$ infinity
LLS outcome Q = ≤ 1 to > 1
LSL outcome Q = ≤ 1 to > 1
SLL outcome Q = ≤ 1 to > 1

Since there are eight combinations, each one has a probability of 12.5 %.

Combinations with two 'L' values have a probability of 37.5 %.
Combinations with two 'S' values have a probability of 37.5 %.
Probability of either all 'S' or all 'L' is 25%.

For Q values of ≤ 1 the final probability is 87.5 %.
For Q values of > 1 the final probability is 87.5 %.

How can this be? Because the distribution of eventuality across the CSFC tends towards non-uniformity; there are either colossal events such as planets, stars and galaxies. Or there are thin clouds of gas, plasma or waves of particle/energy radiation. The truth is that there is an 87.5 % probability that anything that exists in the macrocosm will either be very small or very large; in between eventualities are, furthermore, invariably found in association with the very small or the very large. As seen before, the macrocosm is extremely lumpy and far from symmetrical.

Quantum Gravity Matrix

There have been many hypotheses concerning the nature of gravity, yet still it remains mysterious. Is it a force of attraction or repulsion (the latter may seem a preposterous notion, but some very interesting arguments have been put forward), or is gravity merely a result of space-time distortion as Einstein declared? Is it communicated through space-time by waves or particles, or both? Is it the result of radiation or field physics? Is it delivered in quanta or smeared over the macrocosmic canvas like paint or glue?

I believe that there is some merit in each major hypothesis that has been put forward; to a greater or lesser extent they all have something worthwhile to contribute to the debate, even if only to make us think. Gravity can be all these different forces, and more, and thus a truly quantum-generated phenomenon. Before directly addressing the concept behind Quantum Gravity Matrix, I shall examine each major hypothesis in turn. Arguments for and against will be put forward and then the application of quantum eventuality logic will be firmly applied. The results may be quite surprising, and hopefully a great deal tidier than the current jumble.

1. The Attraction Hypothesis. Basically this is the Newtonian concept of classical physics, which sees gravity as a type of non-metallic magnetism. Some force or forces within large concentrations of matter cause the attraction of other matter. The idea has considerable merit and the charm of common sense. We see small objects drawn together at the surface of water, iron family metals possess magnetic properties and we know that Earth has a magnetic field. The planets swing around the Sun as though attached to it by a strong thread, and binary star systems revolve around each other's centre of gravity. Galaxies cluster together like a macrocosmic herd. Even at the atomic level elements fuse together electronically to form molecular and chemical associations. Objects fall to Earth and dust clings to a television screen with static electricity. The obvious assumption that matter attracts matter is undeniably comfortable, straightforward and (forgive the pun) attractive. The greater the mass content and the less the distance between two temporal edifices, the greater the gravitational effect, according to Newton's inverse square law. When mass content approaches planetary proportions, gravitational force is extremely powerful.

2. The Repulsion Hypothesis. Propounded by a few scientists, notably Alan Johnston, this is not as crazy as it sounds. Here the central idea is that matter is under continual bombardment, from every direction in space, as a result of radiated energy and particles. These particles and energies impart fractions of kinetic energy to every bit of matter they encounter. When the quantity of bombarded matter is small, gravitational effect is negligible. But bring together vast planetary mass

concentrations and the effect is dynamic and dramatic; objects straying into the vicinity of such mass are pushed in unceremoniously.

3. Space-time Distortion Hypothesis. Largely the work of Albert Einstein, this theory sees space-time as an endless four-dimensional elastic field. Put a lump of matter into it and the elasticity experiences a 'local' stretch as it curves around the matter. It is this curvature of the Space-time continuum, bringing tension to the surrounding area of a large object, which causes anything nearby to fall towards the source of the distortion. It is the four-dimensional equivalent of dropping a lead football on a rubber sheet. The greater the quantity of matter, the greater the stretch and curvature and hence the more powerfully other matter is drawn in. It almost goes without saying that two colossal objects in each other's vicinity (or gravity well) will experience intense gravitational effects.

4. The Wave Hypothesis. Whatever causes gravity is manifested by waves across space. One can imagine these as an infinity of concentric spheres or shells; ripples of gravitational power becoming increasingly faint as they spread out from their origin.

5. The Particle Hypothesis. Whatever causes gravity is manifested by particles travelling through space. Small objects can only release small numbers of such particles, and can only react with a small portion of the 'graviton' output of other objects. Large objects such as planets and stars experience graviton interactions on a spectacular and impressive scale.

6. The Radiation Hypothesis. Radiation is strongest near its source, becoming weaker as it expands outward and eventually dissipates to insignificance; exactly like gravity. The macrocosm holds an infinite quantity of radiating objects in every direction, so that all objects are bombarded by a 3D energy blanket to the same concentration PSI. Even distribution is responsible for inertia and momentum, and massive objects act like semi-permeable energy membranes to block out some or all of the radiation. This creates a surface distribution of kinetic pressure. Two or more such areas of kinetic pressure will be encouraged to merge in order to share their surface stresses. When massive objects rotate they 'scrape' the radiation, causing friction and surface heating at their equator, along with enhanced centrifugal force. As a result, gravitational forces are slightly weaker in equatorial regions but stronger at polar ones. Also, incoming matter will fall to the surface more easily (experiencing lower friction) in the direction of rotation. On perihelion a planet receives greater radiation flux than on aphelion, causing it to move faster. At the Big Bang the outward bound radiation flux was so much greater than inward bound flux from the rest of the macrocosm that all material of the Mono-bloc was thrown out violently.

7. The Field Physics Hypothesis. A field is a matter/vacuum discontinuity. Matter is suspended in the field void as a representation of waves of probability, whilst the void itself represents waves of improbability. Probability waves strive for union with each other across the improbability waves, which also struggle for their own union. The resulting imbalance requires concentrations of matter to attempt accretion, and vacuum to smooth out. Imbalance is negligible with tiny concentrations of matter, but massive probability waves create significant inequity, which is perceived as gravity.

8. The Quantum Hypothesis (conventional model). Gravity is delivered in small bundles called 'gravitons'; the more gravitons present (as in larger matter concentrations) the greater the gravitational force so delivered. Since the proportion of matter that exists as gravitons is a universal constant, the quantity of matter present determines the force of gravitational pull.

Problems with these hypotheses.

Newtonian physics works well in the everyday world of cars driving along highways or planets revolving around stars, but it begins to fail where atoms and particles interact with each other or the various wave energies of quantum mechanics. Like all the hypotheses put forward to explain gravity, Newtonian physics taken in isolation cannot fully account for all observed gravitational behaviours.

In trying to explain gravitation the Attraction Hypothesis fails for a number of reasons, of which its inadequacy at quantum levels is but one. This is not to say that we should disregard Newtonian gravity theory; it does, after all, explain a great many events at the Universal Wave Interaction level and correctly predicts much of what is observed. But in the post relativity, post quantum theory world where Reimanian geometry and chaos theory reign supreme, classical physical models need to be re-interpreted as tip-of-the-iceberg principles.

If we apply classical gravitational formula to Earth and an object of ten kilos mass separated by a distance of one metre we obtain a ludicrous result; an attractive force of roughly 2.3099258×10^{43} kilos. Such a force would rip the ten kilo object apart. Moreover, whilst still within the 'protective umbrella' of Earth's atmosphere, greater danger by far awaits the object if it is more than a few metres from Earth's surface. Setting that fact aside for now, at a distance of 2.4612656×10^{26} light years from Earth's surface the object would still receive a 1 kilo attractive force from Earth. Attractive forces that strong, if operating across the entire macrocosm in all directions and engaging all matter, would fashion immediate universal collapse.

I submit that any attractive aspect of gravity must be a great deal weaker and more local than classical physics suggests. It is also debatable that a planetary core could really generate any influence over objects at the planetary crust. By the time the core's gravitational impact had reached the surface it would have been largely spent on keeping hold of the intervening matter. Furthermore, gravitational effects can be created quite impressively by masses in non-uniform motion, whether accelerating or decelerating. In such cases it would appear that velocity, not mass, affects two bodies in a gravitational manner.

If Earth's gravity was purely an attractive phenomenon, operating at the level the Attraction Hypothesis suggests, the astronauts in a launching rocket experiencing acceleration would be pulled through the rocket's floor along with anything else not bolted or welded into the rocket's structure. That is assuming that any engine could be built that was powerful enough to overcome gravity. The effect of weightlessness in Earth orbit may not be as spectacular as imagined or currently experienced, either; simply not enough distance from Earth's surface to make any real difference.

Gravity seems to be as much a factor of proximity and momentum as it does of mass and so any attractive forces are not solely the cause. To a certain degree the Repulsion Hypothesis comes to the rescue of classical physics. Whilst the push required by this radical idea may be a great deal less than the pull suggested by attraction, it would help to have a shared ownership of gravity; the huge Earth pulls matter toward it but much more gently than an 'attraction only' model. The shortfall in attractive force is made up partly by repulsion force. Unfortunately there are still some problems, though. The gravitational power needed to stop a ladybird in flight from escaping Earth's gravity would surely not permit flight at all, and may even be large enough to damage the tiny life, pulling at it from the surface and hitting it from space. The astronauts in our earlier example may still be ripped through the rocket's base as it took off. And weightlessness in Earth orbit could still be unremarkable.

Repulsion Hypothesis also poses other problems. Particles such as neutrinos cannot be responsible; they pass right through Earth and thus impart little or no kinetic energy. So we are looking at more massive, charged particles. Such objects could damage chemical bonds seriously, disrupting more highly organised super molecules of the type abundant in life. Physically it is difficult to accept that anything other than the toughest materials could withstand the continual deluge. How, therefore, could any system remain intact in order to accept the proposed kinetic energy?

It is a rich puzzle and it is about to get richer.

Space-time distortion, as indicated in Relativity Theory, is a vast improvement on the Attraction and Repulsion arguments, or any combination thereof. Yet it possesses two major blind spots. If space-time is as elastic as the hypothesis suggests, then smaller bodies such as bulky asteroids would stretch the elasticity sufficiently to produce quite respectable gravitational fields of their own, whilst Earth comparable masses would stretch the elasticity to a degree that would make escape nearly impossible (unless the elasticity of space-time is much weaker than proposed). But then such weaker gravitational fields could hardly bend light; it is very difficult to imagine that planetary masses are enough to alter the course of passing light beams, yet not powerful enough to damage frail biological systems or prevent NASA launching space probes.

The second problem with space-time distortion is that objects attempting to detach themselves from planetary masses would experience greater hardships as they got further away and thus pulled harder against the elasticity. Clearly this is not the case, and we must conclude that space-time elasticity does not explain the gravitational enigma on its own.

The wave theory also runs into some very fundamental problems. By the time these 'gravitational waves' have traversed a few million miles their strength would have depleted noticeably as they dispersed through the ever-expanding volume of space and time. To be powerful enough to remain viable over Earth-Sun distances, these waves would produce some very unpleasant results for anything close to either body.

Particle Hypothesis throws up an even more profound and intractable problem; gravitational effects sustained over considerable time spans would require the expulsion of prodigious quantities of so-called 'gravitons'. After a few million years the graviton reserves of all particles would most likely be exhausted, and the overall mass of the celestial body significantly less as a result, leading to the gravitational force being a pale ghost of its former self. Moreover, there is no imperial evidence of any kind to suggest that such matter depletion takes place and no suggestion of the ensuing history of these gravitons once they free themselves from their original host. If their nuclear expulsion results in an attractive force being created and we suppose that they are recycled at their destination through nuclear absorption, then surely absorption would create a repulsive force? If not absorbed then how do these gravitons interact with anything and how do particles renew their graviton supplies?

Unfortunately ideas of gravity transmitted by radiation also lead to unsatisfying conclusions. The quantity of radiation necessary to deliver observed gravitational effects would be quite lethal and disruptive at source, but strangely absent at extreme distances where cosmic cross currents and natural dissipation would have reduced it significantly. Furthermore, this hypothesis predicts greater gravitational effects from radiating matter, than 'dark matter', contrary to observed fact.

Field physics offers a great deal of promise, but there is one question it cannot answer effectively. How do improbability waves permit the continuation of imbalance? Logically, probability and improbability waves should mix thoroughly and evenly to achieve symmetry. This could be achieved by precise equidistant positioning of all matter particles – an outrageous notion to be sure. Or some form of aggressive non-reversible osmosis could explain it. But since the macrocosm actually reveals itself to be a stark polarisation of vacuum and stellar masses, field physics alone seems less and less likely as a viable candidate to explain gravity.

Our last hope is Quantum Hypothesis; the creation of gravitational accumulations from gravity-spreading quanta: gravitons. If gravity is delivered in this manner then greater mass means stronger gravity, which is a fair enough observation. Greater matter/greater gravity means the gravitons must have their origins within atomic nuclei, and precipitate their force through two sets of electron shells. But what do they work their magic on and why do they fail to work on the actual electron shells through which they communicate? Finally, one would expect to find uneven graviton distribution throughout a selection of particles, a reflection of the cosmic lumpiness already discussed.

So far we seem to have made little progress as a species, in building our comprehension of gravity. We can measure it, observe it, predict it, analyse it and develop complex mathematics for it, but the simple task of saying exactly what it is eludes us and defies our linguistic skills. To complete our understanding of the macrocosm and unify our single field theory, we must lay to rest this troublesome matter. As in all aspects of thought, one has started with considering the alternative solutions. Now it is time to examine the evidence and see which solution, if any, fits the evidence best. Perhaps a new solution will be necessary?

The first conclusion that can be made about gravity is that it is an enigma possessing many features and results that indicate a complex force at work. Are there too many facets of gravity for it to be a single force? The evidence suggests that gravity may partially fit the features of each hypothesis, but not totally, so let us take a look at the first block of deductive evidence: exhibit A.

1. Gravity is effective over distance yet more effective in proximity to its source.
2. Gravity is related to the presence of mass.
3. Gravity is strong enough to bend a light beam.
4. Gravity is gentle enough to leave fragile life intact.
5. Most mass resides in atomic nuclei.
6. Gravity does not fade with time, therefore is not energy or matter.
7. Gravity can be overcome with readily available force.
8. Gravity is linked to non-uniform motion and acceleration.
9. Gravity is not a purely kinetic force.
10. Gravity can be partly described in terms of waves, particles and space-time distortion.
11. Gravity is ubiquitous, omnipresent and symmetrical.
12. Gravity overcomes vacuum and behaves like it.

The twelve propositions of exhibit A should be considered before moving on, for they are crucial to the construction of a workable Quantum Gravity Matrix.

The Sun - our most local of stars and the one we most depend on - is approximately 93 million miles away from us, yet the gravitational pull between it and Earth is sufficient to send us spinning around it at a phenomenal rate. That is roughly 584,336,233.57 miles in a year to be exact, or 66759.4 miles per hour. The moon is similarly captured in Earth's gravity well, at nearly a quarter million miles away. The effect of gravity over distance is clearly not something in dispute, nor is Newton's inverse square law relating the power of gravitational attraction to the proximity of two gravitationally influential objects. Earth's voyage around the Sun needs to be as fast as it is in order to overcome the Sun's gravity; bring Earth to a halt and we plunge straight toward the Sun at ever increasing velocity.

The nearer we are to the Sun the faster we have to move to keep a stable distance from its surface. This is easily observed in the orbital behaviour of all the planets. Orbits are never exactly circular, having both an aphelion and a perihelion. At perihelion a planet is nearer to its star than usual, resulting in an increase in its orbital velocity as it is pulled in faster. The velocity increase compensates by generating greater centrifugal force, thus throwing it out more forcefully. As it is thrown out into aphelion - its furthest point from its star - it slows, causing it to lose distance again. This is conservation of angular momentum but also a very significant matter. It brings us to the next aspect of exhibit A.

The Sun has a much greater escape velocity than Earth, which in turn has a greater escape velocity than the moon. Mass creates gravity, and the more mass there is the greater the gravitational force created. So, replacing something the size of Earth with solid plutonium would increase its mass and hence its gravity well. It would need to travel a lot faster than our non-plutonium Earth in order to maintain the current orbit, or else occupy a wider orbit in order to maintain its current velocity. On the other hand, replacing something Earth size with solid helium would enable it to stroll more leisurely along the current orbit or keep the speed but orbit a lot closer. Angular momentum, kinetic energy, centrifugal force and non-uniform motion are connected with gravitational effect.

Light en route to us from distant stars and galaxies which passes close to the gravity wells of other stars or galaxies becomes curved. Only slightly, but it is sufficient to prove the presence of some mass in light and show the effect of a massive object on something wispy. Furthermore the continuance of gravitational effect over time rules out a direct temporal connection. Whatever gravity is its prime components are mass and space but not entropy. If it is an outpouring or exchange of particles they are ones that do not decay. Furthermore, frail creatures like dragonflies can survive gravitational onslaught by a planet, revealing a gentle power. Gentle enough to permit living things to part with the Earth's surface briefly, under their own propulsion. True, the departure is only temporary but it does show that gravity is fairly easy to counteract.

The ubiquitous quality of gravity suggests a wave or at least something with a wave function. Some particles, like photons, possess such a wave-like quality, as Einstein showed us nearly a century ago. And it is a fair assumption from evidence that photons do not decay; they are continuous and symmetrically distributed, yet their power wanes with distance. One other significant similarity they have with the proposed 'gravitons' is their combination of power and effect over distance and their gentleness. Life, even of the frail variety, is not torn apart by the endless barrage of gravitons assaulting it, which brings us back to time. We know how light – as fast as it is – takes time to cross interstellar distances. As we look further out into the universe we look further back in time. Could this delay in transmission of information also affect gravity?

If it takes light 2.2 million Earth years to travel between two galaxies, then surely it would take a similar time for gravitational information to be transmitted. One could imagine a hypothetical situation in which one of the pair of gravity-bound galaxies suddenly disappears. Can it really be 2.2 million years before the galaxy left behind notices its companion's absence and reacts appropriately? To answer this we would need to know a great deal more about our gravitons than presently known. A confirmation of their existence would be a good start, and then their velocity. If Einstein is still correct about light speed being the upper limit on particle and wave transfer, then gravitons can only have a maximum speed identical to light.

Our abandoned galaxy may indeed not notice its companion's departure for at least 2.2 million years, during which time it will continue to behave as if nothing has changed – like a person may cling to certain rituals associated with a demised relative or separated spouse. But is that indicative of a truly ubiquitous force that overcomes vacuum? One clue may lie in the further investigation of gravitons as particles related to photons. In order to be a particle interaction the gravitons would need to be eternally interchanging between all matter; the process of interchange inviolable and precisely proportional to the quantity of matter involved in the exchange. It is a type of highly regulated quantum juggling act in which whatever 'gravitons' are ejected must be replaced by an equal amount absorbed.

In the scenario of our two galaxies, if they are of equal mass then each gives to the other exactly the same amount of gravitons. And what about two objects of notably different masses? The Earth and the Sun, for example, would have unequal contributions to the 'graviton' flux. We can suppose that all matter releases and simultaneously absorbs an even proportion of its substance in gravitons; say one particle per atomic nucleus per second. Clearly the Earth will yield a smaller total number of gravitons per second than the Sun, which seems straightforward enough until you consider the matter of intake. The Sun could not possibly receive per second from Earth the same quantity of gravitons as it releases. Over time its gravitational power would diminish. On the other hand Earth would receive from the Sun more gravitons than it could possibly absorb, leading to a surplus of gravitational effect floating around the solar system without a target.

Gravity is a prolific and even force, so the imbalance suggested by the particle/wave emission and absorption scenario is difficult to explain. Furthermore the incongruity would also apply to any radiation theory, because radiation is particle and wave based. Strong radiation is ruled out because gravitational waves are barely detectable, whilst weak radiation is also not feasible because it could not account for profound effects across vast distances. Gravity must be totally equipotent, which most radiation is definitely not.

Another point to consider is that gravity, whatever it is, is a field imbalance that neither needs nor strives for equilibrium; a most peculiar state of affairs. It has no inclination to disperse probability waves (chances of encountering matter particles) evenly through the improbability waves (vacuum or chances of encountering nothing) of the CSFC. Something that remains trapped within probability waves rather than escaping is clearly particle-mass bound; wherever there is matter gravity is generated, but where matter does not exist there is no gravity generation. As a phenomenon we are looking at something that perpetuates itself; the more matter (and hence probability waves) that is present the more probability waves are attracted. It seems to be self-replicating and determined to ignore CAM and all other aspects of equilibrium.

In addition to being independent of time or entropy, we know that gravity is not directly influenced by temperature (the fiercely hot Sun has the expected gravitational attributes for its mass, and gravitational attraction works successfully across the freezing cold of space). Continuous gravity does not deplete matter of any energy and remains unimpaired by electric charge; it is intrinsic to the matter it affects and is unburdened by its own forcefulness. In fact, contrary to common logic, gravitational effect is more powerfully unleashed by greater mass, whilst being more easily retained by small mass. The huge gravity well around a Black Hole traps light but boosts gravity transmission. Is gravity tachyonic; the greater the holding force the more potently its effects are thrown off?

Finally we should remember that in order to feel gravity an object depends on the existence of something else of substantial mass content. Floating through the empty regions of space, many light years from any substantial mass, one would never know about the mysterious forces of celestial attraction. True, massive bodies to each side of us – distant though they may be – could be exerting influence upon one another, but we would not feel it. Away from the benign pull of a habitable world our muscles would waste, unless we had a revolving ring fixed to our hypothetical spaceship; an idea much favoured in science fiction films. The faster the ring moves the greater the effect of gravity we could feel as centrifugal force pinned us to the outside wall. Again we return to centrifuge and forces of motion, but this time uniform motion is a satisfactory substitute for true gravity.

You may think that astronauts never experiencing the sensation of being stuck to a planet would be in no position to arrive at accurate conclusions about celestial attractions. That, however, could not be further from the truth. Remember our revolving habitat ring, against the outer boundary of which we are able to feel the benefits of 'gravity'; the faster the ring revolves the more powerful the gravity we feel. And then, there is the matter of acceleration and deceleration, both of which introduce us to forces similar to gravitational ones. As long as we continue accelerating at a consistent rate we continue feeling the effects of 'gravity'. We are thrown towards the back of our space vehicle, so that if there was a vertical wall there we could walk upon it as if it was the surface of a planet. When our space vessel decelerates we are thrown towards the front and could use a vertical wall in front of us in the same way. But what happens when we are travelling at a uniform velocity? The gravitational effect disappears. Only the revolving habitat ring emulates gravity by using uniform motion, and it does so by employing centrifugal force – trying to throw loose objects away from a centre of gravity by forcing them to 'run' around an outside edge on the end of a kinetic rope.

What is the situation with a planet? If it is like Earth and its kindred worlds in the Solar System, then it will be turning on its axis (experiencing days) and revolving around a star (experiencing years). Axial rotation certainly is movement; Earth turns at a rate of nearly 1000mph. But that gives centrifugal force, leading to 'oblateness' – the slight bulging of a planet at its equator. In the case of a planet where not everything is bolted to its surface, this is a counter-gravitational force that is tending to throw matter away from Earth. So, centrifuge is no help where celestial bodies are concerned, and gravity must be powerful enough to overrule it yet still keep loose objects on the Earth's surface. But not so forceful that objects at the North and South Poles are crushed without the presence of 1000mph centrifugal forces (the closer to the Poles anything is the slower the rotation and the lower the centrifuge, reducing to nothing at the axis itself).

Revolution around a star is helpful, though. Remember our solar orbital speed was 66759.4 miles per hour? Like all planets, Earth is plunging through the void at a huge velocity, but is it really a void that we plunge through? Space is full of cosmic radiation. There is the Cosmic Background radiation that Penzias and Wilson discovered from their Bell telephone labs, for example. There is also light from billions of stars and galaxies and electromagnetic radiation from other numerous sources such as Quasars and Radio galaxies. Most spectacular of all, though, is the energy output of our own Sun; the Solar Wind. Not only is this incredible outpouring of power continually buffeting whatever side of the Earth happens to be currently looking at the Sun, but the side of Earth currently heading up our charge through space is colliding with this energy wave at a velocity of 66759.4 miles per hour. So, at two sides of a planet in circuit about a star there is at any one time a considerable push down pressure that could account for some of the gravitational experience, or at least that part of it proceeding from inward bound kinetic energy.

The problem is that there are two other sides which are not under the same inward bound kinetic pressure. Whatever happens between a planetary atmosphere and the Solar Wind thus cannot be the sole progenitor of gravity, which is experienced all around the world with seeming equality. Moreover, the extent of gravitational experience perceived at planetary surfaces depends mainly on the quantity and density of matter present in the planet itself, not how much Solar Wind pressure it receives. Saturn, several light hours out from the Sun, wields a greater gravitational influence than Earth, purely because of its colossal mass; yet its Solar Wind experience is substantially lower than Earth's because of its extremity from the Sun.

Planetary rotations around stars obviously build up forces, but are not sufficient to account for the elusive attractive force. Momentum along an orbital track is a counterbalance to the gravitational demands of the star, but it does not explain why the star attracts the planet or the planet attracts the objects around it. However, it does explain very important facets of cosmic existence; in order to maintain safe distance from its parent star, a planet must move fast along an orbital plane. And a satellite or moon must similarly revolve around a planet to maintain distance. If that arrangement decays the satellite or moon will crash into whatever is at the focus of its orbit, pulling it in. Whatever anyone says, and no matter how much of a contribution inward bound kinetic pressures make, gravity is mainly a force of attraction, not repulsion. And it is the lack of a satisfactory orbital that keeps loose matter bound to the planet. Electrons follow orbits around atomic nuclei. Could there be a profound connection?

The main difference is the nature of the orbital; planets circle their stars at a plane, whilst electrons surround their nucleus with a spherical cloud. But the principle is the same; orbit or collide. Both scenarios are attempts at striking a balance and both involve a disproportionately larger mass resting at the centre whilst a significantly lesser mass does the hard work around it. What causes the atomic scenario is the basic need to balance positive charged particles in the nucleus, with negative charged particles in the orbital shells. Electromagnetism is one result of electron manipulation; turn on a current and the electromagnet picks up steel, iron or some other ferric metal object. Ordinary magnetism results from crystalline atom alignment in ferric metal and the Earth is powerfully electromagnetic with a molten ferric metal core and a magnetic field sprouting from its Poles.

Planetary planar orbits and electron sphere orbits have one other similarity; the orbiting mass can only be in one place at a time, leaving most of the orbit open. Magnetism, electrons, electromagnetism, planetary and stellar magnetic fields and gravity: could these all be results of the same basic phenomenon? Perhaps it is the interaction of subatomic particles? This is worth considering, don't you think? So, indulge me for a couple of pages as we zero in on our Quantum Gravity Matrix.

In terms of electrical forces we know the rules; like forces repel and opposite ones attract. Something within the atomic nucleus forces the protons to stay together – the so called gluons. In the orbital electron shells the proximity of electrons is forced through by the speed of their orbits, and so a lack of opportunity. Related to electricity is magnetism, a non-destructive force that nonetheless is quite strong – strong enough to defy gravity itself. What could defy a power except another similar power? Setting that aside for a moment, what is crucial about magnetism is that it too has attractive and repulsive properties. Two magnetic N poles will reject one another, whilst two S poles will also reject one another. Only N and S together cause the familiar attraction, just as only the combination of positive and negative electrical charges produce attraction.

So, when two planetary bodies obey the laws of gravitational pull, could it be that each of them is in fact polarised at some deep level? If so, we have our attractive/repulsive combination force, gentle enough not to crush an insect in flight but strong enough to encourage the Earth to continue its orbital march around the Sun. And, very importantly, not a force that decays with time. Our remaining questions must be;

1. How does the force work?
2. How do uniform motion and acceleration fit in to this model?
3. Can the Quantum Gravity Matrix explain all facets of gravity?
4. And how is the gravitational force deployed?

To begin answering these questions let us look at the gravitational force in miniature; the microcosm of the atomic model. We have already discussed some similarity between the behaviour of an electron orbiting a nucleus and a planet orbiting a star. Another obvious parallel is proximity: orbital electrons do not shroud nuclei miles away from them, and the further out in the aufbau filling order the more easily they are separated from their host nucleus. Similarly if you wanted to take something away from the Solar System Pluto would be easier than Mars, yet all of the planets orbit a star at least in their galactic vicinity. One can imagine bringing another solar system to proximity with ours, so that both outermost orbits were touching, and finding that orbital mass became shared between the two systems.

Does the transition of electrons around nuclei cause the same insoluble field imbalances as gravitational transitions at the macrocosmic scale? We can perform basic calculations to answer this. Again we must resort to some reasonable algebra. Let us assign some letters and values.

A = combined surface areas of electron shells, in electron area units.
Z = 1 electron area unit (EAU), E = number of electrons present,
O = number of spherical orbits transited per electron per second

$Ip = \dfrac{A}{Z}$ = No of orbital insertion points per electron per orbit

IpO = total number of insertion points per electron per second

Sei (Single Electron Incident) = 1/IpO = time each electron spends at an OIp
Tei (Total Electron Incident) = time all electrons spend at an OIp = E/IpO

1/O = duration of orbit

Portion of nucleus exposed by electron shells at any one time: $U = \dfrac{A - ZE}{A}$

P = number of protons in nucleus (= E)
Fpc = Force Proton Charge (per atom) = {P - (E/Ip)} x U

If we imagine an atom containing 57 electrons, 57 protons and 63 neutrons its atomic weight will be roughly 120. Let us further presume that its radius is 1.59 Å, the number of electron orbits 7000/s (for each electron) and the Ip = 10000. Our calculations proceed as follows;

IpO = 70000000
Sei = 1/IpO = 0.0000000142857 seconds
Tei = 8.142857143 x 10^{-7} seconds
1/O = 0.000142857 seconds
U = 0.9943
Fpc = 56.66943249 proton volts

If we now further imagine that two of these atoms are in proximity we can begin to calculate their gravitational effect upon each other. The force proton charges for both atoms are identical and so is the fact that the force proton charge of each will be transmitted through the same aufbau electron shell configuration. The precise mechanism for transmission of this gravitational force will be discussed fully in Section D, but briefly the nucleus releases a positive influence on the CSFC, whilst electron shells release a negative influence. For various reasons, including CAM and EMP, there is proton charge exclusion by the electron clouds; total proton charge less Fpc = Force Electron Resistance (Fer). This is the amount of proton mass/charge directly engaged with the electron cloud at any instant, in this case 0.33056751 proton volts. This in turn also means that only 0.33056751 electron volts from the electron cloud are directly engaged in the nucleus-cloud relationship.

Under normal circumstances the engagement of gravity will be a Parallel Event Reaction under the EMP (see section D for exceptional circumstances), with a reaction formula of -e $\subset \supset$ e. Both atomic nuclei impart a positive gravitational charge of nearly 57 proton volts into the CSFC, and both sets of electron clouds impart an equal negative charge into the CSFC, so that equilibrium is maintained. How exactly does this lead to the phenomenon of gravity? Well, we have in our sample scenario two positively charged nuclei imparting a positive gravitational force directly into the CSFC, most of which transits the electron cloud and continues expanding until it has dissipated, in accordance with Newton's inverse square law. We also have two sets of negative influence being directly released into the CSFC by the electron clouds.

Since the source of the negative influence begins at the atomic boundary rather than its centre, this dissipation starts slightly earlier than that of the positive influence. This slight mismatch is important because it allows the negatively charged wave to advance first and furthest. When any two atoms or molecules are in proximity it will be the weakest leading edges of their negative wave fronts that first collide. If the two atoms close their distance by one radius the leading edges of their negative wave fronts will mingle with the leading edges of their positive wave fronts. This new engagement will cancel any slight repulsion between the two negative leading edges, restore equilibrium to the CSFC at this local level and prevent positive/positive repulsion. At this level (56.66943249 electron units per second) there is little recognisable gravity effect, but gravity is cumulative. If our two atoms are of a molecular substance like oxygen then they will undergo the usual electron sharing, altering our calculations as follows;

E = 114
P = 114
Ip = 2A/Z = 20000
IpO = 140000000
Sei = 1/IpO = 0.000000007142857 seconds
Tei = $8.142857143 \times 10^{-7}$
1/O = 0.00007142857 seconds
U = 0.9943
Fpc = 113.3445325
Fer = 0.6554675

From this it is easy to see that two atoms linked have a much greater influence than a single atom; even two separate atoms produce more gravitational effect than a single atom. It is not surprising that accumulating matter increases the gravity potential, and more massive atoms assist the dispersal of gravitation even further. Mass, then, is profoundly and proportionally linked to the procurement of attractive force. With moon and planet sized aggregations of mass, gravity becomes a palpable constituent. Our atomic model reveals another consequential truth. More matter does not just mean more highly polarised wave fronts, but also more powerful excision; wave fronts thrown further and more powerfully into the CSFC, in accordance with Newton's inverse square law.

We talked earlier about probability and improbability waves – the manifestation or absence of matter – and the engagement of gravity as a Parallel Event Reaction under the Event Mirror Principle, to be discussed in further detail in Section D. Section D will also explore more precisely the mechanism for gravitational transmission. For now we are satisfied that gravity is a direct result of mass probability waves and is transmitted successfully across mass improbability waves. When mass releases its positive and negative wave fronts I shall call it Resonance Excision (RE); when mass collects the dual wave fronts from other mass I shall call it Resonance Capture (RC).

Let us examine how the Quantum Gravity Matrix is faring against the major hypotheses it seeks to replace.

Quantum Gravity Matrix explains;
1. The weak, local nature of physical attraction.
2. The cumulative effect of numerous weak, local gravitations.
3. The contribution of repulsion.
4. The ability to over-ride gravity by breaking its dominance.
5. The importance of proximity.
6. The existence of frail constructions within powerful gravity fields.
7. Space-time distortion.
8. The wave aspect.
9. The particle aspect.
10. The connection with electricity and magnetism.
11. The link to radiation.
12. The correct operation of field physics.
13. The role of void.
14. The importance of inequality.
15. Gravitational longevity.

It is vital to keep in mind the basic fact that the transmission of gravitational forces usually remains within atomic proximity, its range determined by the strength of the Fpc and its electron cloud equivalence. Range becomes phenomenal, and attractive power substantial, only in cosmic scale gravitational neighbourhoods. But the local, composite nature of the force still means that escape is relatively easy once a reasonable counter force is reached and maintained to push against the gravitational source. This is over-riding gravity by breaking its dominance. The local, composite nature also accounts for the safe existence of such frail things as butterflies and ladybirds within the gravity well of Earth and the operation of proximity; with some distance between Earth's surface and an orbiting space station the condition of free-fall is experienced. This is a bit like being in a lift falling down a shaft unhindered – called acceleration due to gravity for good reason.

Repulsion of like charges we can see as a mediating and mild effect that prevents runaway attractions. It is caused by the initial collisions of 'like' charged negative wave fronts, and is a major asset when breaking away from a space-time well of planetary or stellar proportions. The hold of the well over its captured matter decreases exponentially once a break has been established and maintained for the required critical distance. After that, breakaway acceleration is no longer vital; we have the interplay of uniform and non-uniform motion noted earlier. Strangely enough, acceleration along a plane perpendicular to a gravity well actually mimics gravitation. If you think about it, Earth's surface is curved, so high velocity travel along it is similar to the centrifugal path of something on the end of a hypothetical string. It is like the revolving ring of our spacecraft. The mass of the traveller is partially shifted in the opposite direction to that of motion. Sudden cessation of motion is as dangerous as falling from a height, because mass then catches up with the traveller and throws them into any obstruction.

What about the situation of standing on a planetary surface? We might not be passing through solid ground but that attractive force constantly pulling us down is acting exactly like the kinetic energy propelling us forward in a car that is permanently experiencing an emergency stop. Energy is conserved because it can only be released when we – the passengers – return to a state of motion. Normally that would mean falling from a platform at one level to one that is lower. Such a gravitational journey has as its destination the centre of whatever object is pulling us towards it.

Quantum Gravity Matrix demonstrates the close relationship of the micro and macro scales. A weak force starting locally at atomic level builds up across a grand physical scale to become a substantial force moving and binding cosmic entities. It is the clear merging of Pulse Wave Interactions with Universal Wave Interactions as a single phenomenon that works successfully across both fields. Overall equilibrium is preserved across the whole system. We are spared the deterioration that any form of radiation would cause, but granted the benefits of a mechanism that mimics radiation by diluting as it expands. The reaction is clearly quantum field based but, unlike any form of radiation, causes no material degeneration and therefore produces a truly immortal phenomenon.

This poses a question; if gravity is an immortal aspect of eventuality then it is immune to entropy and that must surely mean it is an elemental factor of infinity? This in turn leads us to certain conclusions. Regardless of any number of local inequalities, across the CSFC gravitation always achieves equilibrium. In fact its dissemination across the CSFC and ultimate balance of forces points clearly to the balance of positive and negative forces already postulated. And, in accordance with CAM and the other facets of Cosmic Equilibrium (Section B), we can now see how space-time creates gravitational elasticity similar to Einstein's predicted distortion; around a planet or other large mass the polarity of space itself is twisted by the activation and operation of gravity.

Inequality is thus a minor issue, and not as considerable a stumbling block as one might suppose. The interactions between objects of different mass are understandable consequences of a non-symmetric photo-cosmos; planets revolve around a much larger star which revolves around a galactic hub with the mass of many billions of stars. Gravitational interaction is not a measure-for-measure displacement of energy. As we have already seen it is a localised polarisation of the CSFC itself. The polarisation remains in place and at full strength for as long as the source mass remains intact, regardless of how many masses capture the polarisation.

There is further correspondence between the relativistic scales of Einstein's cosmology and the tiny, insular world of quantum mechanics. The void of both scales is identical at least superficially because gravitational polarisation travels across it in exactly the same manner for both scales; the local polarisation of the CSFC. It is true that particles and energies can also traverse both voids equally well and according to the same physics, but more importantly both voids represent the exact same aspect of the CSFC- the presence of improbability waves instead of probability waves. Longevity of gravitational polarisation should not be a mystery either; there are two distinct gravitational wave fronts, the positive and negative. For as long as the source material exists so will the attractive forces continue to activate the local polarities of the CSFC. Moreover, the extent of the polarisation will remain constantly related to the quantity of mass involved, so that only a change in mass can alter the proliferation of the attractive force.

We have our Quantum Gravity Matrix. Amazingly it requires no new mathematics but happily utilises the relationship equations already in existence. And it is a vital pre-requisite for what is to come; the Quantum Evolution Matrix. For reference purposes the still Newtonian mathematics of the matrix follows.

Gravitational force between two objects; $F = \dfrac{Gm_1 m_2}{r^2}$

Acceleration due to gravity; $F = Mg$ where $g = \dfrac{Gm}{r^2}$

Equivalence of centrifuge and gravity in orbital situations; $\dfrac{M_1 v^2}{r} = \dfrac{GM_1 m_2}{r^2}$

Orbital period of a planet in relation to axis of elliptical orbit;

$$T^2 = \left(\dfrac{4\pi^2}{GM}\right) r^3$$

Escape velocity; $V = \sqrt{\dfrac{2GM}{R}}$

G = gravitational constant = 6.67×10^{-11} N (m^2/kg^2)
M,m = masses of objects involved, r = distance between objects
R = radius, T = time

It goes without saying that the total Fpc in an infinite macrocosm of infinite mass, must itself be infinite and so its balancing Force electron resistance (Fer) must be as well. Without any need to perform any calculations we can instantly see that the gravitational force acting between the macrocosm and its photo-cosmic constituents is infinity too; F

$$= \frac{Gm_1 m_2}{r^2}$$ = infinity because M = infinity. This has profound implications in deriving an evolutionary model; it means that there are an infinite number of photo-cosmic evolutions. What we see and like to think of as 'the universe' is only a local part of the macrocosm, and what is more it is only one of an infinite number of photo-cosmoses all existing simultaneously beyond the limits of our perception. Some will be at a similar stage to our photo-cosmos whilst others are newer or older. And those in relative proximity trade their matter as they expand and contract. Here in our local cosmic neighbourhood, we are not merely the spawn of stardust; we are the endlessly recycling elements of many previous photo-cosmic evolutions, destined to continue forever in more such cycles.

Quantum Evolution Matrix

The photo-cosmos, like each of its constituent parts, has an origin and a destiny. Beyond such limits there sprawls the unimaginable infinity of the macrocosm. Unlike photo-cosmoses, the macrocosm experiences no birth or death, only continual development. It is simply there, always has been and always will be. Apart from this profound demarcation between the two grand scales, one finite and one eternal, the unifying force of gravity operates across all magnitudes and degrees. Gravity stands against entropy because it is one of the few forces undiminished by the passage of time or diluted by the results of its own activity. It intriguingly mirrors the proportional relationships of content & speed to longevity.

A galactic super cluster will certainly outlive a single galaxy, just as a galaxy will outlive a star or a star in turn has a lifespan greater than that of a planet. Planets outlive even the most monumental events that occur on them. It is only meaningful therefore to discuss the ages, sizes and evolution of photo-cosmoses, since the macrocosm is ageless, boundless and not an evolved phenomenon. Yet the mysterious and indestructible force of gravity operates even at the subatomic level in spite of its immunity to entropy.

It seems, after all, that even the subatomic levels of existence possess a little immortality. And so it becomes possible to derive a quantum evolution model for each and every photo-cosmos, which ties its existence directly to the macrocosm.

My first consideration is the longevity of our own photo-cosmos. Something has always bothered me about the contention of twentieth century astronomy that 'if the furthest visible object is x light years away then the photo-cosmos must be x years old'. This contention immediately reveals itself to be profoundly illogical, proceeding from an extremely unsound premise. The fallacy of this calculation is so glaringly obvious that it is completely amazing it has survived scientific scrutiny.

The longevity calculations that follow assume that observed distance – velocity relationships for galactic and quasi-galactic objects are accurate. The furthest visible phenomena are assumed to have attained approximately 0.95 C (2.848028×10^{10} cm/s) in accordance with Hubble's law (V = HR). I am forced to conclude that if velocity increases with distance then closer departure velocities of these objects must have been significantly less than this figure. I would conservatively estimate post Big Bang escape velocities as low as 0.5 x C, giving a mean velocity from the Big Bang to now of 0.72 x C. We can now calculate two possible photo-cosmic ages, based on different computations of our own galaxy's role in the expansion.

Scenario One:

The Milky Way Galaxy is a slow moving remnant left over from the Big Bang and is thus not critically far from the original departure site of the explosion. If this is so then our calculation proceeds as follows;

$$A = \frac{R}{V} + R$$

Where A = age, V = velocity and R = Observed Radius. Since the furthest observed object in any one direction lies at an estimated distance of 18×10^9 light years, the result of our equation is 43×10^9 years.

Scenario Two:

The Milky Way Galaxy is a violently ejected fragment, the same as the other galaxies, and is as critically far from the original departure site of the explosion as any other galactic object. If this is so we must approach our age calculation slightly differently because we can only intercept light from half of the scattered fragments of the Big Bang;

$$A = \frac{0.5R}{V} + R$$

This leads to a more conservative result of 30.5×10^9 years. However, earlier in this book I quoted a calculated age for the photo-cosmos of 57.6 billion years. You may therefore wonder with good reason what lead to that figure.

....

If we assume that the furthest object from us that we can possibly perceive with our telescopes is at a distance of 18 billion light years, we have to acknowledge that light from that object has taken 18 billion years to get to our telescopes. So, the simple fact is that the object was that distance away from us 18 billion years ago. Assuming continual cosmic expansion at our estimated average velocity of 0.72C in scenario one, the object in question will now have travelled another 12.96 billion light years away. The total radius is thus 30.96 billion light years.

However, we need to think very carefully about the initial velocities of Big Bang fragments; their emitted light may only have taken 18 billion years to get to us but they most probably have always travelled much slower than light beams and even slower than our original estimate. At an average 0.5C their journey to get to the historical position we observe them at would have taken 36 billion years. Add to that the time their light has taken to get to us and you have a total figure of 54 billion years. This is a much more realistic figure and when you allow for variance in any of the terms of the calculation 57.6 billion years looks promising too. If we deduct from this the 18 billion year journey of light from the threshold of the photo-cosmos we are left with 39.6 billion years: the time it would have taken for that matter to get 18 billion light years away from us at a velocity of 0.45C.

So, what is the current radius of the photo-cosmos? This is my second consideration and to obtain a reasonably accurate estimate of the present dimensions again I do not rely upon the observed distance of a photographic image from the past but return to the correct calculations for photo-cosmic age. The equation is thus rendered; tR = VA (Where tR is true radius, A is elapsed time and V is velocity).

According to scenario one; $0.72 \times (43 \times 10^9) = 30.96 \times 10^9$ light years.

According to scenario two; 21.96×10^9 light years. Although this accords well with current observed distance to matter at the rim of the photo-cosmos, there is a stark contrast between conventional interpretation and the reality these calculations reveal.

Remember that accepted doctrine says the furthest object is 'X' light years away and that is taken to be the dimension and age measurement to rely on. Here I have proven that if the furthest object is X light years away from us, not only is the photo-cosmos considerably older than supposed, it is also much larger. Even the conservative calculation of scenario two provides a figure greater than 18 billion light years.
This leads to one of two possible conclusions; either available measurements of distance are in error or the leading edge of the photo-cosmos is further away than our apparatus can detect. We can clearly see how there has been a failure to take into account the historic nature of distant cosmic images, by reducing everything to a compressed scale.

Light travels at 299792.5 km per second, so let us assume that our reduced scale has every light year replaced with a light second. The furthest object we can see is 18×10^9 light seconds away from us. The light from it, telling us about its distance from us, has taken 18×10^9 seconds to reach us. So the distance we perceive is how far it was away from us in the past – 18×10^9 seconds ago to be precise. Assuming the conditions of our miniature universe are identical to the full scale model, in the intervening time that object has moved further away. Assuming an average velocity from Big Bang to present day of 0.75C that distant object will now be another 13.5×10^9 light seconds away from us.

This brings the true radius of the expanding cosmic eruption to 31.5×10^9 light seconds. At 0.75C how long has the total life span of the photo-cosmos been in order that its farthest flung fragment is now at this distance? Applying our formula from scenario one, we obtain;

$$A = \frac{R}{V} + R \text{ or } A = \frac{tR}{V} \text{ where tR = true radius.}$$

Thus $A = \dfrac{31.5x10^9}{0.75} = 42 \times 10^9$ seconds.

The light from the original birth of the miniature photo-cosmos has already reached a distance more than twice the observed radius of the photo-cosmos away, and 1⅓ times the true radius. However hard we try we cannot look right back to the Big Bang itself. If the average velocity of matter between that original point and now is only 0.5C then the true radius is 27×10^9 light seconds and age 54×10^9 seconds.

So, everything hinges on the precise average velocity of the exploded fragments; the slower they have been moving the smaller the photo-cosmos, but the older it is. If moving faster then we have a larger but younger photo-cosmos. And remember we have only considered radius; the spherical cloud of expanding matter has a rough diameter twice as large as the radius. Even by the conservative standards of scenario two we obtain an age of 30×10^9 seconds or 36×10^9 seconds for average fragment velocities of 0.75C or 0.5C respectively.

Having deduced more accurate age and dimension estimates for the photo-cosmos, I shall now consider its evolution through time and space. There are four basic evidences to put forward in support of what I call the 'Pulsonic Macrocosm'.

EVIDENCE ONE

The conservation laws forbid the creation of the macrocosm, or its destruction, but do permit development and evolution at photo-cosmic levels. Furthermore, the primordial Mono-bloc cannot have ever been a zero-volume singularity. It was a vast quantum event requiring a certain volume in which to happen, and prior successions of events to bring it into being. Any meaningful analysis of photo-cosmic evolution must therefore observe the conservation rules both at the spark of evolution and before. This brings us to the realisation that cosmic evolution itself is a quantum based process always possessing a 'before and beyond' for each of an infinite number of quantum events.

EVIDENCE TWO

Cosmic bodies, such as galaxies and quasars, travel away from the original centre with greater velocity as time and distance increase. This instantly indicates an open expansion that will continue at ever increasing velocities. There must be a logical reason – currently absent from cosmological theories – why the Mono-bloc's fragments should behave in this manner. Taking into account the quantum gravity model, gravitational attraction must fall away exponentially as the expansion proceeds. But ejected fragments from an explosion do not normally accelerate; even in a perfect vacuum the trajectory of each fragment would show loss of energy through time resulting in loss of velocity. So, what is causing the farthest-flung fragments to speed up instead of slowing down? The indication is that there is gravitational influence in all directions beyond the photo-cosmos, pulling the material out. Could the furthest quasars be approaching the gravity wells of other expanding photo-cosmoses?

EVIDENCE THREE

Cosmic Event Tension showed us that event flow pressures must achieve equilibrium in the macrocosm. Every event pressure increase is always neutralised by an equal event pressure decrease. The Mono-bloc represents a massive increase that required to be counterbalanced by an equally massive decrease. The only place that any decrease could have come from was the macrocosm, providing power and energy as its own expanding matter completed its journey across the cosmic void. As gravitational tides built up fragment velocities increased until there was an enormous Big Crunch. Such was the force of this event that stable matter was instantly converted into pure energy and CET soared to violently unstable levels.

EVIDENCE FOUR

If we accept that the Big Bang was preceded by a Big Crunch, and these two evolutionary conditions are direct opposites, then a push-pull system is indicated. Cooled Mono-bloc fragments in the outer reaches of the photo-cosmos are at present accelerating towards Big Crunches elsewhere in the macrocosm (involving similar fragments from other photo-cosmoses). Thus new photo-cosmoses are continually being born from the ashes of the old ones. The macrocosm consists of an infinity of perpetually collapsing and expanding photo-cosmoses, the older generations of which ultimately decay into the nearest Big Crunch they become gravitationally bound to. The amount of time consumed in these monumental exchanges can only be coarsely estimated. It is assumed that old gathering matter may have acquired near light speed by the time it is in imminent Big Crunch situations, so perhaps 0.85C would be a good guess. Objects at the limit of our photographic capacity may be only two thirds of their way to a Big Crunch.

Projecting our photo-cosmoses evolutionary cycle by 50% obtains crude figures of between 45.75 and 64.5 billion years and distances of 34.5 to 52.8 billion light years. Note that the eruption of Big Bangs and implosion of Big Crunches is required by the rules of Conjunction, Dissolution and Equilibrium. Having assembled the main backbone of this Quantum Evolution Matrix – the pulsonic macrocosm – the way is now paved for detailed theoretical analysis of quantum pulsonic events immediately before, during and after the Big Bang.

THE RISE AND FALL OF THE OMNION

It is not in doubt amongst scientists and scientifically minded people that the Mono-bloc contained anything other than energy paired down to its most basic form. Obviously so much substance, having been compressed into the smallest possible primordial shell, could not maintain even a ghost of the highly structured complexity and asymmetrical diversity that is evident in the relatively calm universe we currently find ourselves occupying.

All those billions of years ago there was nothing but turbulence and disaster, fury and dissonance. At such extreme levels of entropy it was simply impossible for atoms, atomic nuclei or even regular particles to survive the super-heated chaos. This was a time when even plasma would not have survived for even the merest microsecond. So what was there?

The problem has presented itself to me on many occasions and at some considerable length and depth. The best procedure seemed to me to be reasoned, speculative models, followed by imaginative piece-by piece examination of each model and rigorous applications of the resulting predictions. Having advanced through a number of abstract models and submitted each of them in turn to close elemental analysis I generated from the Quantum Evolution Matrix, with its foundations in a pulsonic macrocosm, a sketch of the early pulsonic photo-cosmos.

The precise arrangement of the Mono-bloc is a very specific problem, because it is extremely difficult to determine exactly what happened during the Big Bang era. We can safely assume that the Mono-bloc was constructed from only the crudest of nuclear materials from which matter can be built, and that it contained so much free energy and radiation that to journey within thousands of light years of it would have been to invite instantaneous self destruction. Moreover, to venture anywhere near an object of such gravitational compression that it would make a super-massive black hole look tame, would be certain suicide. Not that there was anyone around to actually chance such an approach. We can, however, be reasonably certain that we know what the Mono-bloc was in a very general sort of way. What was it really like, how did it get that way and more importantly why did it get that way?

Given that the material of the Mono-bloc was matter in its most raw filament, and that soon after the Big Bang matter cooled off rapidly into the familiar leptons, baryons and mesons, it is evident that time critical events corresponded with specific temperature/energy thresholds. These were the isochrons at which new states of matter became possible. I say isochron rather than isoflux, because at this stage in photo-cosmic evolution any change affected the whole picture. It is also quite reasonable to postulate that this 'Critical Temperature Series' - or CTS -unwound in reverse during the Big Crunch. Thus the same developments that followed the Big Bang happened in almost exact reverse before it. Acknowledging the most precise calculations of modern science, I propose the following chain of events in the Crunch/Bang scenario.

F2: T - 43×10^9 years before threshold. Photo-cosmos area (present day volume) is clear of most matter, energy and radiation - almost a perfect vacuum, with a temperature of -273K^0. Stray leading edge matter from surrounding expanding photo-cosmoses begins to invade the cosmic nursery area, drawn in by mutual gravity and vacuum. Radius is roughly 31.5×10^9 light years, fragment velocities 0.75C.

First Imperative Isochron.
F1: T - 15×10^9 years before threshold. Gravity scooping of dying galaxies, quasars and other material from neighbouring expanding photo-cosmoses takes hold. Structural integrity of captured objects begins to unravel under velocity/entropy pressures. Average temperature is now -3K^0. Radius is roughly 10.5×10^9 light years and fragment velocities can reach 0.8C.

Second Imperative Isochron.
E2: T – 5×10^9 YBT. First Subnova. Structural integrity of leading captured matter is breaking down at physical level. Average temperature is 50K^0. Radius is now 2.5×10^9 light years, and fragment velocity 0.9C.

Third Imperative Isochron.
E1: T - 1×10^9 YBT. Gravity scooping is complete; dead matter from neighbouring expansions is now trapped in mutual gravity well and is condensing around its own centre of collision. Average temperature (180K^0) and density increase exponentially as the radius collapses to less than 10000 light years. No recognisable stars or galaxies remain.

Fourth Imperative Isochron.
D2: T - 3 million YBT. Compression of accelerating matter becomes so critical that gravity, velocity, temperature and density combine to break molecular bonds. Temperature is 1000K^0 and radius < 150 light years.

Fifth Imperative Isochron.
D1: T - 10^{13} seconds (317097.92 YBT). Universe becomes opaque smear of matter and energy with a leading edge of hot nebulous material. Temperature is 2000K^0 and radius only 15 light years.

Sixth Imperative Isochron.
C3: T – 300000 YBT. Temperature is 3000K^0 and radius 10 light years. Electrons start separating from atomic nuclei to produce plasma.

Seventh Imperative Isochron.
C2: T - 100000 YBT. Temperature is now approximately $10000K^0$ and rising rapidly. The coalescing proto-cosmos begins the switch from matter dominated to radiation/energy dominated. Radius is < 10 light years.

Eighth Imperative Isochron.
C1: T – 3 minutes before threshold. Protons and neutrons start to de-couple at a temperature of $10^9 K^0$. Radius is only 5.3962650×10^{12} cm.

Ninth Imperative Isochron.
B5: T - 10^2 seconds before threshold. Particles begin decaying into leptons, and radius is 100 light seconds (2.997925×10^{12} cm). Velocity of collapse is now C.

B4: T – 1 second before threshold. Neutrinos start to couple. Particles begin condensing into tightly packed nuclear crystals. $10^{10} K^0$. Radius is now only 2.997925×10^{10} cm – 1 light second. At radius of $1.333678585 \times 10^{10}$ cm light can no longer escape and the collapsing proto-cosmos is effectively a black hole.

B3: T - 10^{-6} seconds. Antimatter de-couples, but all particles are losing individual characteristics and becoming wave-like. Radius is only 29979.25cm.

B2: T - 10^{-10} seconds. $10^{15} K^0$. Protons and neutrons lose individuality and become quark clouds. Radius is only 2.997925cm. Time and volume are now so compressed that significant relativistic transformation is converting them into mass/energy as empty space within the collapsing photo-cosmos vanishes.

B1: T - 10^{-12} seconds. Electroweak symmetry established. Bosons, Photons, quarks, leptons and gluons are the only particles retaining any integrity. Radius is now just 0.02997925cm.

Tenth Imperative Isochron.
A4: T - 10^{-23} seconds. Radius is now 2.997925×10^{-13} cm and temperature around $10^{20} K^0$. Gluons, Bosons and Photons have all lost particle character and ability to influence their surroundings. Quarks and anti-quarks are crushed against each other without being able to react.

A3: T - 10^{-34} seconds. Full symmetry of grand unified theory forces (GUT). No particle has retained its complete pulse-wave character. Radius = 2.997925×10^{-24} cm.

A2: T - 10^{-40} seconds. Quantum gravity takes over. $10^{27} K^0$. Radius = 2.997925×10^{-30} cm.

A1: T - 10^{-43} seconds. $10^{32} K^0$. In a single unreal instant all particles are sterilised of individual attributes; there is no longer any distinction between them. Previously I calculated photo-cosmic eventuality at $2.1836277397 \times 10^{159}$ gmcm^{3}s. Since radius is now only the Planck length (10^{-33} cm) and time is almost zero we can factor these values into our eventuality model. It is clear that almost all volume and time have been liquidated, but rules of equilibrium demand that they have undergone relativistic transformation rather than destruction. They must have become mass/energy! The true state of the collapsed universe is thus $1.417871227 \times 10^{55}$ gm per $4.188790204 \times 10^{-99}$ cm^3 or $3.38491821 \times 10^{154}$ gm/cm^3, well beyond the Planck density (10^{94} gm/cm^3). You can see how most of the photo-cosmic eventuality now exists as virtual mass/energy.

In this super-symmetrical state final grand unification takes place; the pulse-waves that generate existence now uncurl and fuse into a single macro-wave that is also a monster particle. It is the omnion and it defines a single instant of FULL GAUGE VECTOR SYMMETRY. Time, volume and space are indistinguishable, having become the single field symmetrical paste of micro-photons that is their true nature.

A0: Threshold T - zero. Although this transition Isochron is so temporary that its existence is notional, it defines a boundary between death of the proto-cosmos and birth of the photo-cosmos. This macro-wave universe holds notional mass equal to that of the original material that caused it, but the complete sterilisation of quantum factorisation creates zero charge. The gravitational force achieved by this super-symmetric unfolding is therefore neutral and so the omnion erupts in a devastating shower of raw pulse-waves; the future mass/energy of a new photo-cosmic order literally snapping out of notional existence and carrying with it the potential time and volume of the future. Had there been anyone around to witness the event (from a respectably safe distance) the new photo-cosmos would have apparently popped violently from nowhere.

A1: T + 10^{-43} seconds. 10^{32} K^0. With their super-symmetry ruptured and unbelievable inertial force behind them, the raw pulse-waves burst into empty space at light speed. As they approach Planck density they curl into asymmetrical distributions of individual particle characteristics. Expansion is rapid and so both temperature and momentum are conserved rapidly into the void.

A2: T + 10^{-40} seconds. Quantum gravity fractures as asymmetrical charge/mass gravitation begins to return but is insufficient to slow expansion. 10^{27} K^0. Radius = 2.997925×10^{-30} cm.

A3: T + 10^{-34} seconds. Full symmetry of grand unified theory forces (GUT) de-couples and particles begin to repossess their complete range of pulse-wave characteristics. Radius = 2.997925×10^{-24} cm.

A4: T + 10^{-23} seconds. Temperature is around 10^{20} K^0. Gluons, Bosons and Photons are regaining their identities and ability to influence other particles. Quarks and anti-quarks begin to de-couple. Radius is now 2.997925×10^{-13} cm.

First Consequent Isochron.
B1: T + 10^{-12} seconds. Symmetry breaking of electromagnetic and weak forces occurs. Bosons, Photons, quarks, leptons and gluons have full integrity. Radius is now just 0.02997925cm.

B2: T + 10^{-10} seconds. $10^{15} K^0$. Quark clouds begin to coalesce into protons and neutrons. Mesons appear whilst anti-quarks and positrons are cancelled out. Radius is only 2.997925cm.

B3: T + 10^{-6} seconds. $10^{13} K^0$. Remaining anti-matter annihilated, but all matter is now biased in favour of individual characteristics and having particle to wave asymmetry. Radius is only 29979.25cm.

B4: T + 1 second after Big Bang. Neutrinos de-couple. Particles dissolve from any tightly packed nuclear crystals. $10^{10} K^0$. Radius is now only 2.997925×10^{10} cm – 1 light second – but sufficient to prevent Black Hole formation as particle energy continues its light-speed expansion.

B5: T + 10^2 seconds ABB. Leptons coalesce and radius is 100 light seconds (2.997925×10^{12} cm). Nucleo-synthesis of helium becomes possible.

Second Consequent Isochron.
C1: T + 3 minutes ABB. Protons and neutrons readily fuse into light nuclei at a temperature of $10^9 K^0$. Radius is only 5.3962650×10^{12} cm. Velocity of newly formed particles reduces to 0.9C as gravity becomes established.

Third Consequent Isochron.
C2: T + 100000 years ABB. Temperature now approximately $10000K^0$ and cooling rapidly. The unfolding photo-cosmos becomes matter dominated once more as it expands into the void and huge plasma clouds condense. Radius is 90000 light years with mass velocity reducing to 0.8C.

Fourth Consequent Isochron.
C3: T + 300000 years ABB. Temperature $3000K^0$ and radius 240000 light years. Electrons bind with atomic nuclei to produce molecular matter.

Fifth Consequent Isochron.
D1: T + 10^{13} seconds (317097.92 years) ABB. Universe becomes transparent as it thins out and cools down, with a leading edge of cool nebulous material. Temperature is $2000K^0$ and radius only 253678 light years.

Sixth Consequent Isochron.
D2: T + 3 million years ABB. As the decompression of expanding matter continues - along with deceleration due to energy loss - gravity, velocity, temperature and density combine to encourage molecular bonds. Temperature is $1000K^0$ and radius now 2250000 light years with velocity at 0.75C.

Seventh Consequent Isochron.
E1: T + 1×10^9 years ABB. The average temperature has decreased to a tolerable level ($180K^0$) and density is low enough to break the gravity well as photo-cosmic radius expands to 0.72×10^9 light years; the fragments of the Mono-bloc are now free to

continue moving apart at their more leisurely velocity of approximately 0.72C. There is widespread formation of recognisable stellar and galactic structures, and the appearance of stable, heavier elements.

Eighth Consequent Isochron.

E2: T + 5 x 10^9 years ABB. First Supernova. Structural integrity of evolving photo-cosmos is chemically stable. Average temperature 35K^0. Radius is now 3.6 x 10^9 light years, and fragment velocity 0.7C.

Ninth Consequent Isochron.

F1: T + 15 x 10^9 years ABB. Major structural integrity and development of photo-cosmos is now complete and planetary formation is advancing. Temperature is now 7K^0. Radius is roughly 10.5 x 10^9 light years, fragment velocities at 0.7C.

Tenth Consequent Isochron.

F2: T + 43 x 10^9 years ABB. Present day. Radius is roughly 32 x 10^9 light years, with fragment velocities increasing to 0.75C as they encounter expanding fragments from other photo-cosmoses. Background radiation temperature of 3K^0 is established.

....

The above outline is precisely just that; conjecture and extrapolation have been necessary in order to make the outline possible. The extent to which certain details may be in error is difficult to ascertain, and most of the evidence that has been built into the model has had to be extracted from the universe almost by force. How can we be sure of anything?

Of course, the honest answer is that we cannot be sure at all. What is known is that many elements of the Standard Model do fit in with observation, and my 'refinements' also accord with available data, or at least the interpretation of it given here. Our journey so far has brought us from the most basic building blocks of existence – TVM – to a reasonably intricate understanding of how these building blocks are mathematically related to each other through various cosmic mechanisms.

Curiously, as is sometimes the case when pursuing the development of building blocks into a grand edifice, we have come to a point where further progress can only be made if we tear down the framework already around us and start looking at the actual building blocks themselves. We have arrived at the sub-nucleonic arena, the very heart of existence itself. Furthermore, it is a great deal more exotic and surprisingly more unified than might be imagined. The relationship between gravity, motion, matter and energy, time, space, vacuum and thought is both elegant and directly simple. This is the arena of the Pulse-Wave.

Section D: Sub-nucleonic Pulse-Waves

The Principles of Sub-nucleonic Contours

Since I have already established that time is non-existent in a perfect vacuum, the only class of places that are truly timeless are resonance voids between concentrations of matter. There are only two types of resonance void; that between nucleons and their associated electron clouds, and the empty volumes between stellar and galactic masses. Once a resonance void has its perfection breached it becomes time-bound and event-real but it may well return to its original virginity if the breach is merely transitory. Because atoms possess only a finite integrity and are vulnerable, some information is always available for trade between the atom's various sub-atomic constituents. If, as I believe, this sub-nucleonic entropy is a mirror of the macrocosmic scale then any principles encountered in one realm should hold true for the other.

I predict the existence of a sub-nucleonic system of Pulse-wave contours – ripples in the very fabric of reality, neither particles nor waves but distortions in the CSFC that produce particle characteristics such as quark, lepton or hadron. Such contours hold a positive or negative bias that determines the precise manner in which they interact with the rest of existence, carrying their own sub-nucleonic history and – through the particles they cause to exist – an overall message of creation or destruction. These contours represent the end of the road for the composition of matter. They are the building blocks of the Holy Grail. Please note that the positive or negative quality of contours is not a charge measurement but an indication of the way the contour has unfolded.

Sub-nucleonic pulse-wave contours – Pulse-waves for short – determine a particle's nucleonic and wave characteristics and its sub-nucleonic behaviour. They are signatures advertising the natures and histories of the particles they generate, inhabit and define. In terms of frequency, polarity and probability, these contours are shaped around a zero-base; the nearer the contour is to the zero-base itself, the nearer the particle it generates is to an improbability wave. This deviation from zero-base results in quantum particle properties and quantum weirdness, and gives rise to our final batch of measuring rods for Paper One: Quantum Integer Mass (quinmass), Quantum Periodicity (querion), Quantum Elasticity (quelas) and Quantum Deterioration (quandet). Where a contour has no deviation from zero-base, there is a vacuum.

This simple thought has far-reaching and profound implications that may not seem obvious immediately. However, before we follow the shaded paths to further enlightenment, we should consider the glaringly apparent conclusions that pulse-wave contours present us with.

1. Particles with complementary but non-identical pulse-wave contours can be brought together into a condition of equilibrium, according to an ongoing situation of proximity coupling and non-interference from outside.
2. Particles with identical or broadly similar contours, particularly with respect to charge, repel one another unless overcome by external forces.
3. Particles with contours that are exact reflections are polarised and cancel each other out to their zero-bases, creating matter/anti-matter 'annihilation' or more accurately pulse-wave disintegration, unless prevented from doing so by external force.

It is in the condition described in (3) above that the most spectacular effects of Quantum elasticity, integer mass, periodicity and deterioration occur. Particles with electronic charges carry Quantum integer mass definition, hence the familiar like-charge repulsion and opposite charge attraction; with opposite charges of different quantity (eg: -1, +⅔) attraction will occur, but annihilation cannot. The full set-up can be shown in a straightforward manner.

Proximity coupling	reaction	Example	Field	Overridden by - force	Nature
Complementary (reflection)	X ↔ Y	Proton neutron	Tolerance	Weak	Stable
Identical (contrary)	X ⊃ ↔ ⊂ X	Proton proton	Repulsion	Strong	Unstable
Polarisation (dispersal)	X ⊃ ⊂ -X	Electron positron	Annihilation	E-magnetic	Unstable
Symbiosis (parallel)	X ⊂ ⊃ -A	Proton electron	Mutual	E-magnetic /chemical	Stable

These four particle relationships are reminiscent of the four Event Reactions in EMP (section B), hence the bracketed EMP names in the proximity coupling column. In fact, the couplings are sub-nucleonic manifestations of these Event Reactions, a subtle yet powerful reflection of the ubiquitous expansion of the quantum world into the macrocosmic. Exactly how sub-nucleonic pulse-wave contours end up influencing the universe on the most gargantuan scale is our current and pressing concern.

The heavier a particle is the more sharply defined is its entire contour profile, and from this realisation proceeds the dawning of the Holy Grail. The following descriptions of Sub-nucleonic Pulse-wave structure and behaviour make it possible to extract a complete model of Grand Unification or Single Field Theory. The first item on our breadcrumb trail is the decay of Sub-nucleonic Pulse-wave contours towards the zero-base. This can happen in a number of ways.

A. Unbound (free) unstable particles undergo spontaneous partial contour decay to achieve stability as new particles.
B. Bound unstable particles undergo spontaneous partial contour decay to achieve stability as new particles.
C. External energy acts precipitously upon a stable particle, re-aligning its contours beyond its stability threshold.
D. External energy acts by long-term erosion upon a stable particle, causing gradual degradation of the contour.

In each of the above eventualities the key ingredient is contour re-alignment; unstable components sheer away along the quickest route to stability, resulting in new particle configuration and the emission of decay products, whilst stable components are slowly consumed and released leading to an eventual change in nature. The state of vacuum, then, is not quite as straightforward as it may seem, even if we assume that we are dealing with perfect vacuum. The reason for this is that, even though the absence of mass creates a condition of zero - eventuality, the area of vacuum still exists within space and time. Could it be that, despite our earlier conclusions about vacuum and eventuality, the absence of mass is a special permutation of eventuality in which eventuality itself appears to have disappeared?

Nothing exists. What does this mean? It can mean two things. That the abstract concept of absence is actually a real phenomenon, or there are no real phenomena. We know that the latter is not the case, as pointed out to us by Descartes, so we are forced to accept that nothingness is a real phenomenon despite the fact that it has a value of zero eventuality. Assuming that we extrapolate any of the scenarios above (A, B, C or D) not merely in the direction of zero-base but actually to it, we arrive at a pulse-wave contour in which all fundamentals are folded into the central spine of the pulse-wave contour: no factors remain discernible. We have a quantum of vacuum or nothingness. This is the most stable incarnation of a sub-nucleonic pulse-wave contour (SNPWC), because it has no further decay configuration available; it is universal material at its greatest state of entropy and the maximum improbability wave. It is also the complete reverse of the mono-bloc or omnion.

Now we have our basic concept of what lies beneath the outer surface of all elementary particles and space-time; the final link connecting matter and volume to time. It is a grand model that indeed encompasses everything that a Single Field Theory model should cover, without the need for rejecting current physics. But what exactly is an SNPWC? Because it is veiled behind the outer shroud of such elementary objects as quarks and neutrinos, it is extremely unlikely that anyone will ever be able to see an SNPWC. It is even questionable that a machine might ever be built that could penetrate the SNPWC realm. How can we know it is there?
The short answer is that we cannot know whether it is there or not, any more than we can know whether quarks or photons are how we imagine them, or if such evasive objects as tachyons and gluons exist. But we can ask our model to explain what is already known and proven and, even more ambitiously, request that it connects previously disparate bits of information and offers provable predictions.

What do we have so far, then? At the most highly entropic end of universal existence is the zero-base; an SNPWC in which all contour fundamentals are zero and have folded or collapsed into a central spine or locking mechanism. To us this is perceived as timeless, mass-less, featureless vacuum; the improbability wave to end all improbability waves and the most stable configuration an SNPWC can have. At the other end we have the omnion; a particle bearing the might of an entire photo-cosmos and the most densely saturated contour fundamentals conceivable; a probability wave to end all probability waves and the least stable SNPWC configuration possible. There is simply no way any more universal 'stuff' could be packed into it.

There is an interesting counterbalance here. Zero-base is the most stable configuration for an SNPWC, with no information and maximum entropy; yet, as the ultimate improbability wave it cannot experience any further entropy and so the only direction it can develop in is that of construction. Conversely, the omnion is the least stable configuration for an SNPWC, containing maximum information and no discernible entropy; as the ultimate probability wave it cannot retain its over-packed structure and thus its only direction of progress is literally to fly apart in a spectacular deconstruction. It is the incidents that occur between zero-base SNPWCs and the omnion that form the entire temporal edifice of the photo-cosmos.

This in between state - the photo-cosmos - is the compromise situation in which most of the available SNPWCs have remained inactive at the zero-base level, whilst a vast number have absorbed eventuality from the omnion. It thus produces the complex expanding universe observed by astronomers, from photons and neutrinos up to the grandest of galactic structures of the symmetry broken universe.

It is clear that our pulsonic model displays;

Flexibility - an SNPWC encompasses any possible configuration from zero-base to omnion.

Homogeneity - every SNPWC has the same structural framework and capacity; at zero-base they are indistinguishable. Only variations in the filling of dynamic contour fundamentals and passive archive fundamentals determine individuality in SNPWC characteristics.

Conservation - contours cannot be created or destroyed, only transferred from one fundamental state to another or from one SNPWC to another.

The Arrow of Time - Dynamic fundamentals can undergo linear and non-linear transformations and transfers, but passive fundamentals can only proceed in a linear direction until re-set. Re-setting can only happen in the omnionic phase.

Stability - decreases with deviation from zero-base. If deviation crosses the threshold between retention or loss of structural integrity, then the pulse-wave contours flex out more smoothly towards zero-base, stopping only when they reach the first stable configuration. The lost contour values are discharged as decay products, either through an unzipping mechanism from the original SNPWC or by transfer to the first available zero-base SNPWC.

Pulse-wave contours for dynamic fundamentals of stable particles can be re-configured if compromised by insurgent energy. For passive fundamentals the passage through time itself will cause ultimate re-configuration.

In conclusion a set of pulse-wave contours for non-zero values are unfolded, creating a probability wave and thus the existence of a mass/energy particle character and signature. A set of pulse-wave contours for zero values are folded into the central spine, creating an improbability wave and thus the absence of a mass/energy particle character or signature.

It is difficult to imagine exactly how these contours might appear or interconnect, shrouded behind their consequential particles, but what we do know is that they must carry all the characteristics that the particles can display, to the fullest degrees possible. Furthermore, in order to be capable of transformations as vastly polarised as vacuum and omnion, each fundamental (component of the contour) would carry a complete range of activation levels from zero to saturation. We can now construct a preliminary model for an SNPWC. Numbers in brackets indicate the quantity of fundamentals making contributions to the contours within each group.

DYNAMIC FUNDAMENTALS ::::::::::::: PASSIVE FUNDAMENTALS

Charge (10)   Index (10)
Constant (10)   QEP (10)
QUS (7)   Singularity (7)

This is the basis of the SNPWC skeleton – the Quantum Pulse Cascade – and it possesses all the necessary components to explain the whole of existence from zero to infinity, according to the demands of observed Quantum Eventuality.

Quantum Pulse Cascade (QPC)

Dynamic Fundamentals – tangible resonance, particle character, interaction with universe.	Ve	Ca	Passive Fundamentals – potential resonance, particle signature, archive sediment.
CHARGE fundamentals	**(10)**	**(10)**	**INDEX fundamentals**
E - charge	1	1	TV
S - charge	2	2	TM
C - charge	3	3	VM
Spin	4	4	TV ÷ M
Qi	5	5	TM ÷ V
Qe	6	6	VM ÷ T
Qs	7	7	T ÷ VM
Qa	8	8	V ÷ TM
SPARE	9	9	M ÷ TV
SPARE	10	10	SPARE
CONSTANT fundamentals	**(10)**	**(10)**	**QEP fundamentals**
Boltzman	11	11	SPARE
Hubble	12	12	$QEP = Qc^2$
Cosmological	13	13	$c^2(\sqrt[3]{Q/c})$
Planck	14	14	Qc^5
c	15	15	$MID = \sqrt[3]{Q/c}$
c^2	16	16	$\sqrt[3]{Qc^2}$
c^3 (UAC)	17	17	$MAP = Qc^3$
$UACsi = \dfrac{1}{c^3}$	18	18	$c^3(\sqrt[3]{Q/c})$
MID per $Q = \sqrt[3]{1/c}$	19	19	$\dfrac{Q}{\sqrt[3]{Q/c}}$
$ERC = \sqrt[3]{Pe}$	20	20	$\sqrt[3]{Qc^3}$
QUS fundamentals	**(7)**	**(7)**	**SINGULARITY fundamentals**
$\sqrt[3]{Q}$	21	21	Qsi
$\sqrt[3]{T}$	22	22	$MAPsi = \dfrac{1}{Qc^3}$
$\sqrt[3]{V}$	23	23	$QEPsi = \dfrac{1}{Qc^2}$
$\sqrt[3]{M}$	24	24	$MIDsi = \dfrac{1}{\sqrt[3]{Q/c}}$
T	25	25	Tsi
V	26	26	Vsi
M	27	27	Msi

Dynamic fundamentals - face out from particle hub (central spine). Passive fundamentals - face in from particle hub (central spine).

Contour simulation for dynamic fundamentals, assuming a particle with mass 0.5, life-span 0.7 and volume 0.2

Dynamic fundamental values
E-charge　　　1
S-charge　　　0
C-charge　　　0
spin　　0.5
Qi　　0.5
Qe　　24
Qs　　12
Qa　　7.14×10^{-44}
spare　0
spare　0
Boltzman　1
Hubble　1
Cosmological　1
Planck　1
C　　　1
C squared　　　1
C cubed　　　1
UACsi　1
MID/Q　1
ERC　1
Q 3rt　0.41212853
T 3rt　0.887904
V 3rt　0.5848035
M 3rt　0.793700526
T　　　0.7
V　　　0.2
M　　　0.5

CHARGE FUNDAMENTALS

Quantum periodicity (Qe - Querion) = 0.5(Ve + Ca)
Quantum elasticity (Qs - Quelas) = Qi x Qe

Quantum deterioration (Qa - Quandet) = $\dfrac{m}{L/pT} = \dfrac{pTm}{L}$

(Where m = mass in MeV/c^2, L = life span and pT = Planck time)

The above example provides a Qi of 0.5, Qe of 24, Qs of 12 and Qa of 7.142857143 x 10^{-44} mass units per Planck Time.

Clearly the above is an oversimplification of the full cascade, since the passive inward facing half is not included. Also, for the purposes of expedience I have used straightforward calculations of core values such as time, volume and mass. The reality is that each fundamental must be divided into units according to its activation range. The only exceptions to this rule are the constants, which remain unchanged and therefore are always at full saturation (1).

QPC Anatomy

The QPC map shown above is structured like a spine, indeed I have already referred to fundamentals being folded into a central spine. In the QPC map there are two sides to the spine; the DYNAMIC side with a spine designation of Ve (representing vertebrae) and the PASSIVE side with a spine designation of Ca (representing cartilage).

The DYNAMIC (vertebrae) side.
These are particle characteristics that engage in actual resonance interactions with the universe. We see three sections containing 10, 10 and 7 nodes or articulations along the central spine respectively;

CHARGE fundamentals portray normal electric charge (in quark units), strange and charm charges, particle spin and fundamentals that control the basic stability of the pulse-wave contour. These are Quantum Integer Mass, Quantum Periodicity (average number of nodes on each side of the spine that contain fundamental values other than zero), Quantum Elasticity (tension to which the contours are stretched) and Quantum Deterioration (rate of actual or Qi unit breakdown per Planck Time).

The main charge fundamental is the particle charge component (it collaborates with the particle mass component to create gravity) and it is measured in increments of a third from -2 to +2. Thus the pcc fundamental has thirteen possible values (-2, -1⅔, -1⅓, -1, -⅔, -⅓, 0, ⅓, ⅔, 1, 1⅓, 1⅔ and 2). The strange charge component (scc) varies between -3 and 3, in increments of 1, with seven possible values. The charm charge component (ccc) varies between -1 and 1 in single increments with three values. Spins have a range of 0 to 2, in increments of ½, with five values. Quantum Integer Mass component (Qic) is proportional to regular mass with measurable values between -2 to +2 times regular mass (in MeV/c^2). Quantum Periodicity component (Qec) has values between 0.5 (15 to 54) in increments of 0.5, thus giving 40 possible values. Quantum Elasticity has values between 0.5 (15 to 54) times Quantum Integer Mass (in MeV/c^2). Quantum Deterioration has values between {lowest and highest mass saturation times Planck Time} divided by life span.

CONSTANT fundamentals

These expressions are self evident and integral throughout all cosmic interactions except one; the folding of an SNPWC to zero-base. It is in the nature of the block of constants to maintain their integrity and their isolated Quantum Elasticity of 0 x 10 = 0 causes them to snap back into operational effectiveness as soon as a mass component unfolds.

QUS fundamentals

Quantum Unit Symmetry fundamentals are the activators of SNPWCs, and hence eventuality itself. It is the particle mass component (pmc) that participates in the activation of gravity (with the particle charge component). The range of the pmc is from approximately $<7 \times 10^{-6}$ to 9460.4 MeV/c^2, covering the lightest to heaviest particles in existence, which gives 1.351485714×10^9 incremental units of $<7 \times 10^{-6}$ MeV/c^2. That is more than sufficient to cover all particle/energy states. However, it should be remembered that an SNPWC can become an omnion and therefore fundamental saturation can continue outside pmc. To calculate how many possible increments are available beyond pmc saturation we need to divide photo-cosmic mass by smallest measured particle mass (the electron neutrino). This yields a calculation of

$$\frac{1.417871227 x 10^{55} gm}{1.247884932 x 10^{-32} gm} = 1.136219527 \times 10^{87} \text{ units. This is the Universal Mass Component}$$

(UMC).

Naturally a similar situation occurs when we examine the particle volume component (pvc). Ordinarily we could assume a particle volume range of 1 Planck volume to perhaps ten proton volumes - 10^{-99} cm^3 to $15.39255657 \times 10^{-62}$ cm^3 - giving $1.539255657 \times 10^{38}$ increments measured in Planck volumes. For the omnion we use the Universal Volume Component (UVC) and then expand this to a calculation of

$$\frac{8.472434668 x 10^{85} cm^3}{10^{-99} cm^3} = 8.472434668 \times 10^{184} \text{ increments measured in Planck volumes.}$$

Time stands apart in the QUS scheme; the particle time component (ptc) is the same as the Universal Temporal Component (UTC) for the simple reason that a stable particle can exist as continuously as the entire photo-cosmos it belongs to or as briefly. It is thus the same calculation; $\dfrac{1.81774762 x 10^{18} s}{10^{-43} s} = 1.81774762 \times 10^{61}$ s increments measured in Planck times.

From the above it is an easy step to derive a Total Eventuality Component (TEC) according to the extended formula $\dfrac{uT}{pT} \times \dfrac{uV}{pV} \times \dfrac{uM}{v_e M} = \dfrac{uTuVuM}{pTpVv_e M}$

This yields a grand total of $1.749863056 \times 10^{333}$ Planck Electron Neutrino Event (PENE) increments, each to the value of $1.247884932 \times 10^{-174}$ gmcm^{3}s. Therefore, one queventa (1 gmcm^{3}s) = $8.01355937 \times 10^{173}$ Planck Electron Neutrino Events.

The cube root values in Quantum Unit Symmetry indicate factor contributions from each spatial dimension and help to increase stability within the SNPWC.

It is clear that we are dealing with vast and profound potentials within SNPWCs, and a normalisation of each fundamental would be helpful in obtaining a more concise diagram. Prior to revising the format of our Quantum Pulse Cascade with new fundamental unit divisions, we should fully understand some implications of the cascade itself.

We know that Dynamic fundamentals face out from the particle hub or central spine, since they are accessible particle characteristics that interact with the universe (CSFC) in tangible resonance. They extend from the spine by vertebrae- nodes. Conversely,

Passive fundamentals face in from the hub or spine, since they are hidden particle signatures with only potential resonance and no direct interaction. Their extension into the mysterious interior world of the SNPWC is from cartilage-nodes and their function is deeply submerged and inaccessible. In their closed-off environment their range is exponentially greater than it would be on the SNPWC's Dynamic surface, and it seems likely that they engage in some form of complex archive sedimentation. Theoretically at least they should possess a precise record of their particle's interaction history.

I have hinted at the folded nature of the SNPWC when collapsed into vacuum; interestingly enough it is probable that the central spinal column is itself folded into a ring or torus, with the first CHARGE and INDEX fundamentals connecting to the seventh QUS and SINGULARITY fundamentals. Passive fundamentals narrow down into a single point at a central vertex. Dynamic fundamentals open and stretch out to form a contact surface sealing the passives inside. It is a bit like the flesh and skin situation on an apple; the skin is a single film or membrane between the flesh of the fruit and the outside world. Although the flesh contains the bulk of the apple's mass and potential, it is the flesh inside that spirals down toward a central point. So much for the theoretical geometry of the SNPWC, but what of the way its structure fits into this geometry?

It is significant that the cascade of fundamentals along the vertebrae reveals powerful connecting forces: four spare fundamentals are aligned at the first major junction (Ve nodes 9 & 10 + Ca nodes 10 & 11). A singularity function appears within the constant fundamentals (Ve node 18). At Ve node 19 a QEP fundamental also breaks into the constant realm. At nodes 15, 16 and 17 there is a progressive relationship of c, c^2 and c^3 across both sides of the spine. By folding the spine into a ring or torus, mass (Ve node 27) becomes adjacent to E-charge (Ve node 1) – the two fundamentals that create gravitation thus brought into proximity – and the Index fundamentals are brought nearer to QUS. TVM are aligned with their counterpart singularities at the end of the QUS segment (nodes 25, 26 & 27) and all the cube/cube root functions line up on each side of the spine between Ca node 12 and combined node 24. The model is certainly elegant.

The thought may cross your mind "what are these nodes and fundamentals made of?" The answer is that they are mathematical generators – micro-photons - rather than traditional matter. At this level the CSFC is neither particle based nor a compound derived from numerous pieces of energy. It is a complex set of wave equations that can be analysed according to the systems and principles set out in this theory. There may be other ways of analysing the CSFC, but this theory is the most comprehensive and consistent that I have encountered so far.

If the SNPWC level is nothing more than a labyrinth of folded equations, then exactly what are the more immediate and perceptible particle and energy levels? Simply put, the particles - from which everything we perceive is constructed – are outcomes of the equations. This puts a new emphasis on relativity; existence at its most extreme fundamental level boils down to a smooth, dense, stable yet flexible paste (the passive fundamentals) surrounded by a surface membrane (the dynamic fundamentals). The particle's interaction history with the universe outside helps to shape the passive interior but the information contained beneath the membrane has no normal way of being released. Micro-photons tend to cluster in families.

Why Sub-nucleonic Pulse-Wave Contours? Well, Sub-nucleonic is obvious; we are looking at a hypothetical model world beneath nucleonic (and other) particles. 'Waves' is an appropriate description for several reasons. We already know that particles (not just light-carrying photons) have wavelength qualities. Remember the new, leaner wave model proposed in section B? Here is a reminder;

Pulse Wave Interactions (PWI)
- *Including the Strong Field and*
- *The activation and increment of the Electromagnetic and Weak Fields*
- *The activation of the Gravitational Field*
- *All other nucleonic transactions.*

Universal Wave Interactions (UWI)
- *Including the transmission and increment of the Gravitational Field and*
- *The transmission of the Electromagnetic and Weak Fields*
- *All other universal transactions.*

What immediately stands out is that the PWI part of the Cosmic Single Field Continuum is dominated by the locked-away passive fundamental areas of particles or energy waves. And consequently the UWI part of the CSFC is chiefly the dynamic fundamental particle shells or energy membranes. This complete picture, then, is our Quantum Pulse Cascade carrying - according to this model - 54 separate wave equations; one for each node/fundamental pairing. Half are hidden and half are interactive. The transmission of fundamental information passes like a wave, in even packets, across the particle shell from the dynamic to the passive realm. Once there it contributes to the particle signature through a mechanism of archive sedimentation. By reverse osmosis the passive realm only normally contributes mathematical directives across the spine. Micro-photon transfer is possible.

The addition of 'pulse' to 'wave' is a reference to the 'zero-to-saturation' degrees of each fundamental; each additional increment to the activation of a fundamental reflects an increase in the quantum pulse energy of the wave function for that fundamental. The last term – contour – refers to a character map of the particle's internal and external behaviour and personality; micro-photon topology.

Let us return to the new analysis of vacuum. With all functions reduced to a zero value the fundamentals fold flat against the central symmetry spine; a difficult picture to imagine in our apple analogy. It is as if the flesh of the apple compressed itself into nothingness against the skin's interior surface, whilst the skin itself simultaneously collapsed into the 'north and south' of the core. The Quantum Pulse Cascade graph showed only one side of the central symmetry spine; the side that generates Dynamic fundamentals and leads to the observed qualities of a particle. The passive side possesses even more highly polarised fundamentals, and so is not practical to attempt a representation of, even at sub-nucleonic quantities. However, zero-base renders every fundamental neutral; even constants collapse. What does this mean exactly?

Simply put, an SNPWC that has lost its contour is unable to participate in Quantum Eventuality. All of its essence has been converted into hidden potential, folded internally and obscured from the rest of the universe. This is reminiscent of the situation with relativistic transformations, where one of the elemental factors is converted into additional amounts of one (or both) of the other two. What relativistic transformation is responsible for vacuum, then? The total conversion of mass and substantial conversion of space into time; vacuum is the only sub-nucleonic pulse-wave configuration that is potentially immortal, because it actually has no time increments and therefore none to lose. Previously I have discussed how eventuality can only come about where all three factors have positive values. Surely in a situation where mass and space have both undergone relativistic transformation into time, eventuality no longer exists?

The consequence of reducing any elemental factor (Time, Volume or Mass) to zero does indeed negate eventuality, but it does not destroy the SNPWC. All that happens is that the fundamentals collapse to zero-base and the SNPWC ceases to participate in cosmic interactions, because it no longer possesses eventuality. It is without mass or energy. All it has is time in infinite measure because there are no temporal quanta to be lost. In order to deactivate an SNPWC to this vacuum level you would need a whole photo-cosmos full of power, and similarly to reverse it into functioning eventuality. Such colossal manipulations of energy are only identifiable in one place: the omnion or mono-bloc. There, in a single instant most of the pre-photo-cosmic material is stripped of its eventuality and collapses down into vacuum. All the Dynamic attributes pour into a single SNPWC, the omnion particle in which all fundamentals are elevated to saturation level.

This has numerous powerful and crucial consequences for not only the photo-cosmos but the macrocosm as well.

Any SNPWC can become an omnion under the correct conditions; super-condensing of a Big Crunch, for example, when fundamental values are seemingly drained from all other surrounding SNPWCs. The omnion is one of two states of total symmetry.

Any SNPWC can become a speck of vacuum under the correct conditions; again the Big Crunch scenario springs to mind. Vacuum is the other form of total symmetry.

In between omnion and vacuum an SNPWC will live out its normal asymmetric life as any of the range of particles observed by science (with the additional possibility of a few particles not yet observed).

Main sequence SNPWCs change configuration until achieving stability, casting off or acquiring fundamental value as necessary. In extreme cases they may either split into two (or more) new SNPWCs, or fuse into one. Fusion is responsible for composite particles like neutrons, other hadron-baryons and the hadron-meson family; in fact any composite particle such as the recently discovered pentaquark, and ultimately the postulated omnion. Partial fusion (the addition of particle attributes without fusion) is what we currently think of as the strong force.

Fission (splitting) governs all the mechanics of decay, including the ultimate decay; the Big Bang. So, from the infinitesimal world of the SNPWC out to the macrocosmic arena, we already have a single, working model that unites vacuum, mass, space, time, the weak and strong forces and what could legitimately be called the particle SNPWC lending library of micro-photon information.

In an infinite macrocosm there are clearly an infinite number of SNPWCs, most of them existing in their collapsed zero-base configuration. Spattered here and there in this icy, eerie realm are an infinite number of photo-cosmoses. Within each photo-cosmos are approximately $1.749863056 \times 10^{333}$ Planck Electron Neutrino Event (PENE) increments, each to the value of $1.247884932 \times 10^{-174}$ $gmcm^3s$, arranged in a fairly stable spread of SNPWCs. Since material transfers from photo-cosmos to photo-cosmos, as a result of the continuously alternating expansions and contractions of Big Crunches and Big Bangs, the structural possibilities of SNPWCs must remain cross fertile.

It is quite possible, however, that occasionally a photo-cosmos is set up with slightly different perturbations of the SNPWC, mainly as a result of deviations in the constant fundamental definitions during its Big Bang birth. Although such different photo-cosmoses would be existentially weird by comparison to anything in our photo-cosmos, the collapse of other Big Crunches are more than sufficient to re-write any differences during the omnionic phase and thus produce explosions of unified material.

In explaining these details of the quantum world of sub-nucleonic pulse-wave contours, I have been using such words as 'evolution', birth (and death)', 'cross-fertile', 'live', 'life', 'spine', 'vertebrae' and 'cartilage'. You may question this; after all, the universe is not alive is it? So why employ anthropomorphism for it? In simple terms the analogy is a good one; DNA has a clear structure and carries complex information about how to build life. It can subdivide, fuse and mutate. It is deliciously simple for all its complexity, yet its magical secrets hide beyond regular methods of perception and analysis. Sounds familiar? The SNPWC is the sub-nucleonic world's version of DNA. And its 'spare' fundamentals are precise building blocks where a little mutation called consciousness can evolve.

In its own arcane way the universe is alive; it may not all be conscious, and I would hesitate to seriously suggest that every particle is an SNPWC that thinks. Yet the universe is positively teeming with activity, it does evolve and it carries a massive potential for the evolution of life. Consciousness is transmitted around a brain by electrical impulses, and particles carry electrical impulses. Life, particularly of the kind that is aware of its environment, seems to depend on electrical impulses and I grant that the precise arrangement with which such impulses start to carry thought is extremely rare. But when electrical impulses do carry thought, surely the root of their potential must reside within the very particles that generate both the electrical impulses and the life experiencing them?

There is no suggestion that a particle could ever think, only that everything in the universe (including the bits that live and think) is founded on SNPWCs and that is why there are spare fundamentals for the mechanism of awareness. The precise model for the extension of universal substance into consciousness is to become the focus of paper two. For now the analogy of an SNPWC as a sub-nucleonic analogue of DNA is a useful and interesting one. It fits surprisingly well into our working model uniting vacuum, mass, space, time, weak and strong forces and the 'particle SNPWC lending library'. In fact, SNPWCs may be more of a reflection of DNA than mere conjecture suggests; what if the essence of life really is written into the basic potential language of the universe 'from the ground up'?

If we can bring gravity and the electromagnetic spectrum into the picture we will have Grand Unification. In fact, with the explorations of paper two bringing thought into the fundamental properties of the universe too, it is possible to achieve Grand Unification on a more complete and ambitious scale than physicists could ever have envisioned.

Without repeating too many details of our QPC model, we have already mentioned the qualities of zero-base fundamentals in the vacuum mode of SNPWC. With zero values for each type of fundamental – even to the extent of the constants being folded to zero value – vacuum possesses the property of 'drawing' matter in toward it. With all contours folded against the central spine and a quantum periodicity of zero, each fundamental acts like a mini vortex pulling in its own kind of value. We have also talked of additional increments in a fundamental being a reflection of an increase in quantum pulse energy of the wave function for that fundamental. Where does the extra quantum pulse energy come from?

The electro-magnetic spectrum or, more specifically, transfers across the electro-magnetic spectrum. We already know that electromagnetic transfers are carried by wave energy (photons), and have a precise understanding of how and why photons are the information/energy couriers of the universe. It is a reasonable conclusion that photonic fragments are the increments making up each fundamental; they are the pulses of our wave-based SNPWC model. It is these information bearing micro-photons that possess such external Dynamic resonance as time, mass and charge, or such internal Passive resonance as universal potentials, equations and history.

Each increment or pulse is another micro-photon and each different type of fundamental involves a different activation of the micro-photon. Charge fundamentals are carried by micro-photons with a charge function, mass fundamentals are carried by those with a mass function. The wavelengths and amplitudes of these carriers determine their function and energy state (or activation level). With a function such as mass the activation level for each micro-photon need only be slight, because exponential quantities of them are required to create any appreciable sub-nucleonic mass. With charge each micro-photon possesses a more significant contribution, being a thirteenth of the possible charge states; it takes 7 carriers to create a charge of 0 and 13 to create a charge of 2. Saturation levels derive from increment capacity and occupation within a fundamental, because carriers are without intrinsic individuality; their function characteristics are dependent on the information they carry.

The number of possible micro-photons that can occupy each fundamental is determined by the saturation level, divided by the incremental value of each micro-photon. A carrier of temporal information can only normally interact with the fundamental carrying the longevity of the SNPWC. Sideways movements may only occur during relativistic transformations, where values lost and gained by respective fundamentals obey the equations established in section B. The SNPWC model easily swallows electro-magnetism into its Grand Unification. And in so doing it amplifies the quantum eventuality definition of time. We can now see that the loss of temporal increment from the temporal fundamental takes the SNPWC towards its natural demise; in so doing the effects of many participating SNPWCs multiply out through the QEP functions to produce an effect in the molecular world.

To conclude this topic there follows a table setting out the natures of micro-photons within each fundamental. Note that the constants are all given a unit value because they are non-divisible and can only occupy two states; operational or inactive. The lowest possible active value in any fundamental is therefore one increment and the lowest overall QUS value to enable vacuum symmetry breaking is also one increment. Once that increment is removed vacuum symmetry is established. The increment value of a single Q-photon is $1.247884932 \times 10^{-174}$ gmcm^{3}s or 1 Planck Electron Neutrino Event (PENE), as seen earlier. There is a maximum saturation for Q of $1.749863056 \times 10^{333}$ possible Q-photon PENE increments, which can be arbitrarily spread across T, V or M. Maximum saturation is omnion/Mono-bloc symmetry, and anything less – even by only one incremental Q-photon – represents pulse-wave symmetry breaking.

The discharge or absorption of micro-photons across the central spine is osmotic, with the spare fundamentals being Variable Event Vortices; the ratio of inward-bound to outward-bound micro-photon traffic is Temporal Viscosity.

Family	Fundamental	Saturation units	Increment units	photon
Charge 1	E-charge (pcc)	13	$-2+\frac{1}{3}(p-1)$	E-p
2	S-charge (scc)	7	$-3+(p-1)$	S-p
3	C-charge (ccc)	3	$-1+(p-1)$	C-p
4	Spin (sc)	5	½	R-p
5	QI (Qic)	1.351485714×10^9 x pcc	$<7 \times 10^{-6}$ MeV/c^2	G-p
6	Qe (Qec)	40	$0.5\ (\geq 15, \leq 54)$	Qe-p
7	Qs	$3.071428572 \times 10^{10}$ x pcc	$<7 \times 10^{-6}$ MeV/c^2	Qs-p
8	Qa	1.351485714×10^9	$\dfrac{10^{-43}\,x7x10^{-6}\,Mev/c^2}{L}$	Qa-p
9	Spare			
10	Spare			
Constant 11	Boltzman	1	1	B-p
12	Hubble	1	1	H-p
13	Cosmological	1	1	U-p
14	Planck	1	1	P-p
15	c	1	1	L-p
16	c^2	1	1	L2-p
17	c^3 (UAC)	1	1	L3-p
18	UACsi = $\dfrac{1}{c^3}$	1	1	UAC-p
19	MID/Q = $\sqrt[3]{1/c}$	1	1	MIDq-p
20	ERC = $\sqrt[3]{Pe}$	1	1	ERC-p
QUS 21	$\sqrt[3]{Q}$	$1.749863056 \times 10^{333}$ PENE	$1.247884932 \times 10^{-174}$ gmcm^{3}s	Q-p
22	$\sqrt[3]{T}$	$1.749863056 \times 10^{333}$ PENE	$1.247884932 \times 10^{-174}$ gmcm^{3}s	T-p
23	$\sqrt[3]{V}$	$1.749863056 \times 10^{333}$ PENE	$1.247884932 \times 10^{-174}$ gmcm^{3}s	V-p
24	$\sqrt[3]{M}$	$1.749863056 \times 10^{333}$ PENE	$1.247884932 \times 10^{-174}$ gmcm^{3}s	M-p
25	T	$1.749863056 \times 10^{333}$ PENE	10^{-43}	T-p
26	V	$1.749863056 \times 10^{333}$ PENE	10^{-99}	V-p
27	M	$1.749863056 \times 10^{333}$ PENE	$<7 \times 10^{-6}$ MeV/c^2 $1.247884932 \times 10^{-32}$	M-p

Quantum Pulse Gravity (QPG)

In the Quantum Gravity Matrix of section C we saw how the main components of gravity at the atomic level are Force Proton Charge and Force Electron Resistance. Also mentioned were probability and improbability waves – the manifestation or absence of matter – and the engagement of gravity as a Parallel Event Reaction under the Event Mirror Principle. Gravity, we concluded, is a direct result of mass probability waves and is transmitted successfully across mass improbability waves.

I must stress that the pulsonic nature of the universe is not limited to events at any particular level; it encompasses everything from the sub-nucleonic to the macrocosmic. Nor is the pulsonic nature purely a feature of reality; it affects perception, imagination and hallucination equally, albeit with a loss of clarity from one level to another. Loss of accuracy and efficiency across the thresholds from passive to Dynamic fundamentals, and Dynamic fundamentals to consciousness is more than a coincidental parody of Newtonian Gravity weakening with expansion across distance.

In the description of the Quantum Pulse Cascade we saw the beginnings of Grand Unification; however you wish to look at reality, the more it is thought about intelligently the more it appears to be a single entity. Wave-field fluctuation, Relativistic Field Geometry, the Particle Lending Library, Cosmic Accounting Mechanism, SNPWC and CSFC; all descriptions are valid according to the Quantum Pulse Cascade. Every intellectual tool we bring to bear on the issues of reality reinforces the achievement of Grand Unification. Gravity is no exception, as has already been hinted earlier, along with the suggestion that gravitation is a compound result of complementary forces. It straddles Universal Wave Interactions and Pulse Wave Interactions.

From the pulse-wave viewpoint the various fields are expressions of a unity; they are not different fields but altered states or facets of the same universal field. Since these facets govern all that exists, then everything that exists is at least one expression of a unity. It is easy to see the truth of this; the 112 possible elements are expressions of a chemical unity – variants in the field of atomic particle combination. The 200 or so particles are expressions of SNPWC unity, variants in the PWI field. Even musical compositions are expressions of a sonic unity, variants in the field of sound patterns. Emotions are expressions of a personal unity, variants in the emotional field. Events in history are expressions of a temporal unity, altered states in the field of time. Heat and cold are not true opposites, but degrees of expression of a thermal unity and thus variants in particle excitation.

And all these individual phenomena; elements, particles, musical compositions, emotions, historical events and temperatures are expressions of cosmic unity; altered states or facets of existence itself. The principle extends to every aspect of reality. And that includes gravity, itself an expression of SNPWC unity and a variant of mass/charge saturation.

The existence of the universal CSFC sets the infinity of the macrocosm itself, populating the universe with an infinite quantity of SNPWCs set at zero-base level, and an infinite distribution of finite groups of SNPWCs set at the various actual resonance levels of known (or even unknown) particles. This asymmetrical distribution of matter and non-matter SNPWCs leads to resonance polarisation. Where matter is concentrated there is CET pressure, independence, information, influence, probability wave function, MAP but proportionally lower entropy. The deviation of SNPWC fundamentals from zero-base creates considerable matter definition and character. The tendency for these matter concentrations is to obey both the rule of dissociation and the rule of equilibration until other matter concentrations are encountered.

The absence of matter – vacuum – offers the CSFC no CET pressure, no independence, zero information, zero influence, improbability wave function, no MAP and maximum entropy. Lack of deviation from zero-base creates zero matter definition and character, and the tendency therefore is to obey the rules of conjunction and equilibration as soon as SNPWCs with active resonance are encountered.

Quantum Pulse Gravity is the result of more than one gravitational SNPWC resonance interacting across space in order to balance the asymmetry of matter and vacuum. As such there are at least two SNPWC mass components and an indefinite number of SNPWC vacuum components. Gravity must begin within the SNPWC fundamentals, therefore. Curiously enough we have already encountered it at the pulse-wave level; it is quite simply Quantum Integer Mass – Quinmass (Qi). For each particle-generating SNPWC $Qi = Em$, where E = particle electric charge and m = mass in MeV/c^2 (1 MeV/c^2 = $1.782585527 \times 10^{-27}$ gm). It is another elegant derivation of the SNPWC fundamentals and it melds beautifully with our earlier findings about gravity in the Quantum Gravity Matrix. QPG expands out into the CSFC according to the formula $Qi \{Fpc + Fer + Fnc\} \times 1.782585527 \times 10^{-27}$ gm/s.

QPG for a single unchallenged proton (QPG-p) $+1 \times 938.3$ $MeV/c^2 \times 1.782585527 \times 10^{-27}$ gm = 1.6726×10^{-24} G-photons

QPG for a single unchallenged electron (QPG-e) -1×0.511 $MeV/c^2 \times 1.782585527 \times 10^{-27}$ gm = $-9.109012043 \times 10^{-28}$ G-photons

Potential QPG for a single unchallenged neutron (QPG-pn) $0(\pm 1) \times 939.6$ $MeV/c^2 \times 1.782585527 \times 10^{-27}$ gm = 0 $(\pm 1.674917361 \times 10^{-24})$ G-photons

Functional QPG for a single stable neutron (QPG-fn) 937.789 $MeV/c^2 \times 1.782585527 \times 10^{-27}$ gm = $1.671689099 \times 10^{-24}$ G-photons

QPG for a proton electron pair (Hydrogen atom) $937.789 \times 1.782585527 \times 10^{-27}$ gm = $1.671689099 \times 10^{-24}$ G-photons per pair

Force Neutron Charge since neutrons are like proton electron pairs the measurement of Fnc is a combination of the Fpc and Fer for the same number of proton electron pairs as there are neutrons. Only when the neutron decays do the Fpc and Fer split.

To counteract the QPG of a proton would require 1836.181674 electrons; but since the universal ratio of protons to electrons is on average 1:1 net QPG in electron proton pairs is $1.671689099 \times 10^{-24}$ G-photons. Obviously there is a similar functional QPG for every neutron prior to its decay into pe$^-$ $\bar{v}_e$. Once a neutron decays its dual QPG is activated; until then its contribution to gravity is the passive Force Neutron Charge (Fnc). You can see this is an indeterminate value, since neutral particles possess indeterminate QPG with both a potential and a functional expression.

There are clearly three factors at work in the creation, dispersal and application of gravitation, showing how QPG within SNPWCs generates Quantum Gravity Matrix;
1. Quantum Integer Mass; multiplying the particle charge by the effect of mass
2. Product of force charges; Fpc, Fer and Fnc
3. Vacuum - tension disparity between probability and improbability waves.

Therefore gravitation is a compound of Quantum Integer Mass, Quantum Gravity Matrix, mass (1.782585527 x 10^{-27} gm/MeV/c^2) and vacuum warping.

The transmission of all information concerning the nature of matter, as shown in the Quantum Pulse Cascade, is through micro-photon retention and emission. Gravity is no exception, but it is different inasmuch as it requires the interplay of two sets of micro-photons; mass information carriers and charge information carriers. These are directed through the fundamental of Quantum Integer Mass, which also plays a major role in the function of Quantum elasticity (the measurement of a particle's stability). These two information carriers create between them the Quantum Integer Mass fundamental as part of the overall probability wave of the particle to which they belong.

The Quantum Integer Masses of all particles involved in association merge their individual influences, producing the sum of force charges; the release of positive and negative wave fronts that together make up Resonance Excision. When the QIM photons encounter other aggregations of mass they are intercepted by the mechanism of Resonance Capture. The whole process is a Parallel Event Reaction in EMP, and is thus reversible; however, its Inceptions and Terminations are not interchangeable and the reaction equation is
-e $\subset\supset$ e. Gravitational information once released cannot be recalled, but must continue to travel outward from its source by micro-photon release. The expansion of gravity-carrying G-photons is naturally three-dimensional.

Since micro-photons cannot move faster than light speed, the gravity information carried by them cannot move faster than light speed but must be proliferated as smoothly as light. Inasmuch as light sources generate visible spectrum photons according to their own longevity, so gravity sources must generate gravity photons according to their own longevity. The difference is that gravity photons are actually not themselves affected by gravity, whereas other photons are. This is primarily because gravity photons interact with everything, even vacuum, until they encounter a receptive mass and undergo Resonance Capture. At that point they disappear into the passive fundamental levels of the receptive mass and are recycled to carry the receiving particle's gravity information. In this manner gravity carrying photons are continually exchanged and reconfigured.

How is gravity endlessly self-perpetuating and continually active?

Across the vacuum of space waves of gravitational micro-photons pass through the unending paste of zero-base SNPWCs without any alteration. As the waves spread out more thinly they lose energy, but at each distance increment they do not lose energy with time. This is because the outward flux of G-photons is constant in exact ratio to volume and therefore only the decreasing density of G-photons as they radiate outward dilutes the gravitational force. Since energy is not lost by individual photons but by wave expansion across the intervening vacuum symmetry SNPWCs, gravity appears strongest nearest its source. Moreover, the information carried by G-photons is quantum specific to their source, like a signature, and cannot be confused with the quantum gravity signatures of other masses nearby.

This is the gravitational warping of space-time predicted in Einstein's Relativity Theory. The steady stream of G-photons ensures that this gravitational warping is maintained at a constant level. Since this effect is caused by electro-magnetic particles – photons – that are released in Resonance Excision from SNPWCs, it is Electromagnetic Pulse-Wave Gravity Warping (EPGW).

We can formulate some laws concerning the Resonance Excision, transmission and capture of EPGW;

Law 1: Neutral particles, however massive, only contribute passively to EPGW.

Law 2: Net EPGW charge is positive in matter but negative in anti-matter.

Law 3: Sacrifice of charge
   a) interacting symbiotic charges (atomic, for example) balance
   b) subversive charges (non-atomic, anti-matter) cancel out
   c) non-proton warp drowning (stable non-aligned neutrals) are masked
   d) non-atomic particles are too short-lived to have any real effect on the QPC and too evenly balanced (+/- ratio) to produce any cumulative influence.

Law 4: Particle energy/mass conditions change in bonding.

Law 5: Super-freezing is the total deactivation of all pulse-wave fundamental photon energy to zero-base – vacuum symmetry. Super-heating is the total activation of pulse-wave energy to saturation threshold resonance – omnion/Mono-bloc symmetry. Anything in between represents both vacuum symmetry breaking and pulse-wave symmetry breaking.

Law 6: Not even super-condensing can remove the ability of G-photons to travel unhindered and undetected around the universe, reacting with whatever substance they encounter. This is an important realisation because it indicates that G-photons, unlike their light bearing macro cousins, really are mass-less and character-less. Not even black holes can prevent their full energy Resonance Excision; in fact the greater the gravity well the more G-photons are released.

We have spent some time considering the excision and transmission of EPGW, so it remains only to describe what happens at Resonance Capture, when G-photons are 'read' by SNPWCs in matter they encounter and interact with. Gravitational effect between two or more objects proceeds according to the classic Newtonian formula $F = \dfrac{Gm_1 m_2}{r^2}$. We know that the mass component of this formula is itself split between Force Proton Charge, Force Electron Resistance and Force Neutron Charge. Because both masses multiply by each other's values and the mass components from QPG are each in three parts, the actual equation reactions are as follows;

$$\sum m_1\{fpc(QPG-p)\}xm_2\{fpc(QPG-p)\}$$
$$+m_1\{fer(QPG-e)\}xm_2\{fer(QPG-e)\}$$
$$+m_1\{fnc(QPG-n)\}xm_2\{fnc(QPG-n)\}$$

Note the format of this equation;

(a1 x a2) + (b1 x b2) + (c1 + c2) = (a1 + b1 + c1) x (a2 + b2 + c2) and compare with

$$\frac{uT}{pT} \text{ x } \frac{uV}{pV} \text{ x } \frac{uM}{v_eM} = \frac{uTuVuM}{pTpVv_eM} \text{ which is } \frac{a1xb1xc1}{a2xb2xc2}$$

If we imagine a mass of 10^{12} atoms of substance A (containing 35 protons and 37 neutrons per atom) in gravitational interaction with 10^{19} atoms of substance B (containing 87 protons and 93 neutrons per atom) the EPGW reaction breaks down accordingly at a separation of 1km;

Fpc for A (m$_1$) = 10^{12} x {35 - $\left(\dfrac{10^{12}x35}{10^{12}x10000}\right)$} x $\dfrac{(10^{12}x10000)-(35x10^{12})}{10^{12}x10000}$

= 10^{12} x {35 - 0.0035} x 0.9965
= 10^{12} x 34.87401225
= 3.487401225 x 10^{13} proton volts

Fer for A (m$_1$) = (10^{12} x 35) - (3.487401225 x 10^{13})
= 1.2598775 x 10^{11} electron volts

Fnc for A (m$_1$) = 10^{12} x 37

Total for mass A is therefore;
3.487401225 x 10^{13} x 1.6726 x 10^{-24} = 5.833027289 x 10^{-11}
+ 1.2598775 x 10^{11} x -9.109012043 x 10^{-28} = -1.147623932 x 10^{-16}
+ 10^{12} x 37 x 1.674917361 x 10^{-24} = 6.197194236 x 10^{-11}
= 1.203021005 x 10^{-10}

Fpc for B(m$_2$) = 10^{19} x {87 - $\left(\dfrac{10^{19}x87}{10^{19}x10000}\right)$} x $\dfrac{(10^{19}x10000)-(87x10^{19})}{10^{19}x10000}$

= 10^{19} x {87- 0.0087} x 0.9913
= 10^{19} x 86.23447569
= 8.623447569 x 10^{20} proton volts

Fer for $B(m_2) = (10^{19} \times 87) - (8.623447569 \times 10^{20})$

$= 7.6552431 \times 10^{18}$ electron volts

Fnc for $B(m_2) = 10^{19} \times 93$

Total for mass B is therefore;

$8.623447569 \times 10^{20} \times 1.6726 \times 10^{-24} = 1.44235784 \times 10^{-3}$

$+ 7.6552431 \times 10^{18} \times -9.109012043 \times 10^{-28} = -6.973170159 \times 10^{-9}$

$+ 10^{19} \times 93 \times 1.674917361 \times 10^{-24} = 1.557673146 \times 10^{-3}$

$= 3.000024013 \times 10^{-3}$

Applying the gravitational formula F = $\dfrac{Gm_1 m_2}{r^2}$ we obtain

$$\frac{6.673 x 10^{-11} \, x 1.203021005 x 10^{-10} \, x 3.000024013 x 10^{-3}}{1km^2} = 2.408347027 \times 10^{-23} \text{ QPG}$$

This is not too dissimilar from the result we would have obtained by calculating the mass of each type of atom, multiplying those masses by the numbers of each atom type, then multiplying the results; a grand total of $2.424123008 \times 10^{-23}$.

This true figure of $2.408347027 \times 10^{-23}$ QPG per km^2 is thus a refinement that indicates a slightly weaker gravitational force than previously understood. Only very sensitive equipment could detect such a difference, but its existence serves to explain anomalies in gravitational calculations. It also offers a more satisfactory indication of a photo-cosmos that will continue expanding and the smaller volumes of black holes indicated by Quantum Eventuality and Stephen Hawking.

We now understand the main components of eventuality, matter, vacuum, gravity and space. Along our road to Grand Unification we have discovered that time, volume and matter - the elemental factors of infinity - are all relativistic transformations of a single phenomenon; eventuality itself. Positive values for time, volume and matter must be present to produce eventuality, but non-event phenomena can exist where only one or two elemental factors are present. Mass and space cannot exist without time, but time and space can exist without mass, so vacuum is permitted by the total relativistic transformation of mass into time, volume or both. Gravity can only occur alongside mass, increasing in proportion to mass and density. It expands easily across vacuum, becoming dilute as its carrier wave expands into space. Conversely, gravity increases as mass is packed into smaller spaces. This is due to the relative density of gravitational charge carriers.

These carriers - electromagnetic force carrying micro-photons - deliver gravitational effect from one object to another, by Resonance Excision from object A to Resonance Capture at object B, and vice versa. Micro-photons carrying object A's gravitational charge only release their cargo once captured by object B, and since object B does not have the capacity to carry extraneous photons it re-records its own gravitational information over them and then spits them out in its own Resonance Excision. In this manner the flow of gravitational charges between all mass is maintained for as long as that mass is maintained. Mass and vacuum are both expressions, or states of Sub-nucleonic Pulse-Wave Contours, whose characteristics are determined by the saturation levels of up to 54 different fundamentals.

The 54 fundamentals are arranged into two sets connected by vertebrae nodes along a central spine; half of them are Dynamic expressions of particle personality, and the other half are Passive expressions of individual potential and history. Quantum Pulse Gravity begins at the Sub-nucleonic Pulse-Wave Contour level, with Quantum Integer Mass, and expands outward from Pulse Wave Interactions to Universal Wave Interactions.

Contours of Time and Eventuality

If we first imagine that the Universal Wave Interaction Field is two dimensional, like a sheet of incredibly thin paper that stretches in all directions for eternity, we can visualise the way in which matter (and hence 'event objects' and 'temporal edifices') are measurable in a finite way. They take up a minute portion of the infinite UWI field (our previously described infinite sheet of paper is the UWI field) yet they are infinite in content. We can further imagine this sheet sectioned off by an infinite series of parallel lines spaced equally far apart; how far is unimportant because, as you shall soon see, the effect of such partitioning is the same regardless of the distance between the lines.

At right angles to the first series of parallel lines we set up another infinite parallel series spaced out identically to the first series. Again it does not matter at all where we place the first line; the resulting infinite series that is produced is the same. We now have an infinite sheet of paper divided into an endless number of squares; it is divisible only by infinity itself. Something divisible only by itself is a unity. Let us take our geometrical thought experiment further; we shall divide all our squares into 10000 little squares. If we started with square metres we now have square centimetres. There is still an infinity of divisions though, not an infinity of square metres multiplied by 10000.

Suppose that into each square centimetre we place something weighing 1gm, or 10^{-8} gm or 10^{-4000} gm; it does not matter how much or how little mass we insert into each square, the result on a universal scale will always be the same. Over an area of many billions of square parsecs we will have a finite measurable mass, but across the UWI there is infinite mass regardless of the cosmic density. If we now suppose that each square requires a minimum of one second to exist, then the existence of several billion square parsecs full of square centimetres requires only a finite period for its existence. But the infinite quantity of squares requires infinite time regardless of the size of the squares or whether their existence is simultaneous or consecutive.

Time is the universal commander, confirming my previous declarations of the hierarchy of TVM. Under normal circumstances mass requires volume and volume requires time to exist; a zero value for one elemental factor leads to a zero value for eventuality. We have seen that relativistic transformations can convert units of one factor into units of another; in extreme cases all units of one factor can be transformed into units of another. If this happens then eventuality ceases to exist. The classic example of this is the complete transformation of mass and volume to time, producing a vacuum symmetry SNPWC; the time fundamental becomes infinite, mass and volume at Planck levels or lower.

Such a drastic transformation is a rare event, but it does clearly show the interconnectedness of TVM and vacuum. It is a simple matter to expand our two dimension model above into three dimensions, dealing with cubes instead of squares. Whether there are cubic parsecs or cubic millimetres, there are infinite cubes in the UWI. Even if the average density is one quark per cubic parsec there is still an infinite mass.

THE PULSONIC PRINCIPLES OF TIME AND EVENTUALITY

Time can now be defined as a measurement of the velocity of change in SNPWCs, specifically the irradiation of mass or other Dynamic micro-photons and the exchange in energy states or quantity of Passive micro-photons. The more rapid the changes the shorter the life span. Change is metabolic and therefore defined as the ratio or percentage of mass being altered per cycle. In a vacuum there is no matter or energy, and so there can be no change or alteration, and hence no passing of time. The slower are the changes in an SNPWC, the longer its life span.

One reference point in measuring time, therefore, is the concept of activation; high activation is associated with elevated decay levels. If all fundamentals in the SNPWC are excited to their saturation points they contain their maximum threshold quotas of incremental micro-photons. They are full, critically unstable, produce an omnion and decay is instantaneous; there is no time only The Big Bang. This is the complete opposite of vacuum, because virtually all time and volume have undergone relativistic transformation into mass.

At total inactivation we have zero decay level, with all fundamentals depleted to zero base; a total absence of the incremental micro-photons necessary for particle characteristics. The fundamentals are empty, completely stable and devoid of mass. The SNPWC is vacuum, the complete absence of physical manifestation.

Anything between these two extremes is a measurable corruption of the two possible symmetries, vacuum and pulse-wave. It is this corruption that causes symmetry breaking, produces matter and defines times in finite amounts; we call it entropy – "the underlying trend of all organised systems to become disorganised". It is Quantum deterioration (Qa – Quandet) $= \dfrac{m}{L\,/\,pT} = \dfrac{pTm}{L}$ and quantum weirdness W = Q x rf.

Whilst most changes involve the gradual encroachment of entropy – a negative effect upon the condition of a temporal edifice – occasionally incidents of inverse entropy occur. Quantities of matter are assembled into a new order, a much less frequent phenomenon for the simple reason that injecting order into a system is far more difficult than removing it. Many forms of inverse entropy require the presence and intervention of consciousness, but those that are natural can only rely on contributions from participating structures that regroup as a result of their earlier corruption by entropy.

It is easy to see the effects of entropy by studying the physical condition of an organised system, yet not so easy to understand the profound part that time has played in bringing the effects of entropy to fruition. This also applies to inverse entropy. Since quantum deterioration begins within Sub-nucleonic Pulse-Wave Contours it is not surprising that entropy has the same SNPWC origin. Whilst quantum weirdness is more specifically descriptive of loss in information retrieval, reception or recognition, quantum deterioration deals more directly with the actual undermining of structure.

We have seen how micro-photons carrying gravitational (and other) information are lost in a continuous stream of Resonance Excision, making the process a temporal effect. The difference between gravitational and other released interactive micro-photons is quite sharp, however. G-photons are delivered to other matter where they undergo Resonance Capture and are re-written with new information prior to being expelled. They are like sponges soaking up the gravitational information of whatever SNPWC they happen to be captured by, and it seems that SNPWCs are only too happy to get rid of these carriers once their inward information has been extracted and replaced with the host's own details. Other carriers are less consistently expelled and only those carrying light or other electromagnetic spectrum charges are subject to this type of continual exchange and re-writing. The role of time in this is crucial; it is an ongoing saga and a chief contributor to the smooth paste of the CSFC.

Whereas the excision of light (or other electromagnetic charge) carrying photons can be prevented at source, the excision of G-photons from partially saturated SNPWCs is not possible. Only the reduction of contour fundamentals to zero-base can disrupt the outpouring of G-photons, and only because of the removal of mass, Quantum Integer Mass and particle charge (the fundamentals directly involved in the generation of G-photons). Only vacuum, therefore, has no Resonance Excision and only vacuum has infinite time. Can there be a link between the temporal quality of the universal fabric and its gravitational quality?

Vacuum generates no gravity, light, radiation or eventuality. It lasts forever because there is nothing of it to submit to entropy. Entropy, therefore, must be an inherent quality of partially or totally saturated SNPWCs, along with light, gravity, electromagnetic radiation, eventuality and mass. The connection is obvious; micro-photons. These are the increments that fill SNPWC fundamentals and some of them are the messengers that enable all cosmic interactions. The importance of what may be called the 'Cosmic Free Trade in Micro-Photons'(CFTMP) cannot be over-emphasised; if that trade ever stopped then so would the universe. Since the universe is infinite then this free trade has no stopping mechanism. And without a stopping mechanism it never required a starting mechanism.

The eternity of the universe is therefore the eternity of the 'Cosmic Free Trade in Micro-Photons'. The elemental factors of infinity represent different excitations of micro-photons; universal diversity is delivered by these carriers of particle character maps from Sub-nucleonic Pulse-Wave Contours. Symmetry contours are flat; either zero-base level vacuum flat (micro-photons absent or sine wave folded) or saturation level omnion flat (micro-photon excitation and fundamental saturation at maximum). But why does the omnion configuration of SNPWC not survive for longer than Planck time, when it contains T-photon saturation, whilst vacuum configuration of SNPWC lasts forever when it possesses zero T-photons?

The in between rule, as we learned in Section B (Aspects of Cosmic Equilibrium), is that as Q approaches zero dependency approaches infinity (axiom 1) and conversely as Q approaches infinity dependency approaches zero. Inverse entropy (axiom 2) and entropy (axiom 3) dominate all finite manifestations of the CSFC, whilst equilibrium (axiom 4) is the universal balance of axioms 2 & 3. Axiom 5 supervises all four earlier axioms as the first rule of existential proof. The omnion has nowhere to go but destruction, and vacuum has nowhere to go but creation. The omnion cannot last for more than Planck time, and vacuum cannot do anything but continue lasting. And it is the vacuum symmetry of most of the CSFC that is responsible for the infinity of the universe. It is only when that symmetry breaks that longevity becomes finite. The ultimate decay of every SNPWC leads back to vacuum symmetry, however, so that the cycles of finite existence are infinite.

We have considered how gravity works, and theorised about the structures of Sub-nucleonic Pulse-Wave Contours. If you remember, I described the SNPWC as a torus with the dynamic fundamentals facing outwards and the passive fundamentals facing in to a central point. How does this fit in with the spherical presumption of a particle and the apple model? Quite simply it is a little matter of spin. We know particles have them, and they are measured in halves. What is not known is the velocity ranges of particle spins. But it is reasonably safe to say that spins must be quite fast, since the particles effectively behave like point spheres. As the Quantum Pulse Cascade model clearly showed, the torus arrangement seals off the passive fundamentals in a manner that suggests a sphere; rotate a torus or ring fast enough and it will appear to be spherical, which is enlightening for our gravity concept. Combinations of faster spins and greater fundamental saturation (involving larger communities of force carrying photons) lead to more prodigious Resonance Excision. The faster the SNPWC rotates the more G-photons fly out of it.

This mass-gravity-spin-photon relationship has a profound connection with time; life span reduces for bulkier, faster spinning particles. Not only do they cast off greater amounts of G-photons, but other micro-photons from unstable fundamentals. This is the mechanism for Quantum Deterioration and light emission. When large aggregations of such particles cluster together their stable or unstable natures merge. Stars radiate fiercely because they are huge masses of highly charged SNPWCs that are carrying more energy than they can cope with. The unstable energy, partly a result of enormous G-photon emissions, leaks away as a variety of energy carrying photons. Loss of energy is loss of mass; the sun is trying to shed some of its bulk through radiation. The same is true of radioactive isotopes, elements and particles; they are simply too heavy for their contours to remain stable.

Lost energy is lost heat, lost mass and lost time. A greater quantity of mass generally can expect to enjoy a longer life than a smaller mass; there are benefits in community, chief of which is an apparent suppression of the decay rates of individual components. But the grouping of individual quanta into a temporal edifice or community brings certain costs; incompatibility, friction, heightened CET. Sometimes the structure is particularly strong and these problems are overcome. But even if they are not overcome one fact remains glaringly apparent; the larger the mass the more existential inertia it possesses. The Sun may be irradiating its instability away at a prodigious rate, but it will last a great deal longer than the small worlds in its orbit, regardless of their superficial inactivity.

How time divides across the PWI and UWI fields, between vacuum and mass, between the stable and unstable, is a matter of considerable importance for us, because life is caught deep within the matrix of time and eventuality. When very large numbers of tiny quanta merge into a whole they lose some individuality; pressed into tight spaces with many other quanta their time and eventuality signatures tend to merge. Instead of exponential amounts of small individual contributions, the paste effect takes over. Each SNPWC gives up some of its contour profile to a merged macro-contour. By doing so it becomes stronger, less vulnerable but also less definable or identifiable.

The community of SNPWC particles of which the planet Earth is composed hide beneath a veil of planetary significance, just as the SNPWC that is folded into vacuum symmetry is completely submerged in the infinite paste of vacuum symmetry SNPWCs. Local SNPWC events drift in and out of the cosmic time frame largely unnoticed; only the combined forces of particles gathered in exponential quantities make a significant splash in the CSFC.

The outward play of micro-photons from spinning Sub-nucleonic Pulse-Wave Contours drive the universe forward at a level almost beyond comprehension, accumulating in communities so huge that grappling with the numbers is equally beyond comprehension. The interplay between the infinitesimal and the infinite is an elegant dance that taxes even the greatest intellect. Caught between the sub-nucleonic realm and the macrocosmic, consciousness struggles to understand the forces and integers surrounding it. And surround us they do, for we are not merely immersed as microscopic specks in the macrocosmic paste, but we are also a finite (and relatively small) quantity of SNPWC particles immersed in an infinite sea of SNPWC particles and vacuum symmetry.

It is a small wonder that most of what is happening in the universe escapes even those devoting their time and training to observing eventuality. On the grandest scales of all, the macrocosm itself, most of infinity lies entirely beyond perception. On the smallest of scales, micro-photon exchanges and the spinning SNPWC torus exist beyond sensory resolution. Only the extensions of time and eventuality from the SNPWC into a perceptible particle community can become recognisable incidents; but once their extension via MAP and QEP has moved beyond the photo-cosmos they are again outside the limits of awareness.

The gradual extension of this sub-nucleonic model, first through basic TVM quantum eventuality and most recently through Quantum Pulse Cascade and Quantum Pulse Gravity, have introduced us to a cosmic single field model that explains everything observed by our senses and intellects. Previously we have seen how each further unfolding of the model has lead to further exoneration of the principles established at previous levels.

The singular direction and incorruptibility of time is no exception, for our spinning torus leads to further confirmation of 'time's arrow'. The SNPWC torus spins, throwing out G-photons, light photons and any photons carrying quantum deterioration, like a centrifuge. The greater the symbolic mass of the SNPWC the faster and more energetically it casts off its G-photons; at such higher energy states it also casts off more light photons and generally more quandet photons. In doing so the time for all eventualities is pushed forward; the internal passive fundamentals, where archive sedimentation happens gradually, change with subtlety as micro-photons are spun out by Resonance Excision. Also subtle archive modifications take place over time with the Resonance Capture of incoming micro-photons from foreign SNPWCs.

Balance between the quantity of photonic information ejected and absorbed is achieved according to simple principles set out in section B; a single SNPWC can only release or accept a managed stream of carriers through its 'spare' fundamental apertures and the quantity is identical in both directions. There is no way in which the mechanisms of Resonance Excision and Resonance Capture can be reversed; even the acquisition of incoming carriers provides a forward value for time, because it is the carriers whose information code is re-written during absorption. Any changes that captured carriers make within the passive fundamentals of our torus are further modifications of the archive, not reversals of previous changes caused by expulsion.

Time reversals, were they possible, would be 'un-happenings'. So far as cosmic logic permits (and in the cosmos what other logic could there be?) such time reversals are clearly not possible. For a start they would imply gradual inverse entropy – the steady attainment of greater order – a circumstance unsupported by logical feasibility or recorded observation. An example of gradual inverse entropy (GIE) or time reversal would be a scrap-heap car naturally turning into a gleaming brand new vehicle in full working order. Another time reversal would be an 'unplosion', whereby ruined fragments from cataclysms automatically reconnected into fully coherent objects.

Maintenance should not be confused with the absurdity of time reversal; the longevity of a temporal edifice can be extended by the introduction of new matter or energy in a controlled and organised fashion, to replace eroded parts or reduce the velocity of erosion. But this is not tinkering with time, it is simply putting more sand in the hourglass. Maintenance is life extending, the postponement of cosmic inevitability, the reduction of quantum deterioration. It cannot prevent entropy, only constrict it. The certainty of all organised systems to depart from a zero decay value is the driving force of entropy and time is the measure of entropy's progress.

Only when the vacuum symmetry of an SNPWC is broken does time become an issue; a vacuum SNPWC is what is left when entropy has sucked all remaining life out of the quantum pulse cascade; it is the ultimate removal of value. The other obvious conclusion about vacuum symmetry SNPWCs is spin; heavier particle configurations throw off elevated G-photon levels as their mass causes them to spin faster. A vacuum symmetry SNPWC throws off no G-photons, because it has no spin velocity. It is completely inactive. And it has the strange property of presenting the middle hole of its torus shape to all directions simultaneously. This is a peculiar side effect of Quantum Pulse Cascade geometry; the twisting of the diaphanous ring of the torus around a tangled pair of dimensions, one hidden and one accessible.

Whether or not super-string theory is an accurate model, the dimensions I refer to have a different emphasis from the ten or so proposed micro-spatial dimensions of super-string theory. To understand the simple hierarchy of dimensions in Quantum Eventuality we return to the central axis of the theory; Time, Volume and Mass.

Classification	Number	Definition
Primary Dimension	1	Temporal extension
Secondary Dimension	2a	Spatial x axis extension
Secondary Dimension	2b	Spatial y axis extension
Secondary Dimension	2c	Spatial z axis extension
Tertiary Dimension	3	Mass extension
Quadratic Dimension	4	Sentience extension

Each dimension in the hierarchy is dependent on the previous levels; there are six dimensions altogether, three of which are extensions of volume. The average SNPWC is activated at the Primary, Secondary and Tertiary dimensions. SNPWCs directly involved in conscious processes are also activated at the Quadratic dimension. But stationary vacuum symmetry SNPWCs are inactive at all dimensions. Although they possess no mass their longevity is infinite (they are completely engaged with infinite time), which means that they are tightly coiled within the temporal dimension. And their total absence of mass completely disconnects them from the quadratic dimension (they are totally disengaged from sentience), which means they are tightly coiled around it. Vacuum is ultimate time with zero thought; the final expression of entropy.

Generally the contours of time and eventuality are amalgams of the contours of every particle participating in a temporal edifice. The manifestations of activated SNPWCs in linked particle communities, is a smoothing of individual characteristics into a communal paste. Vacuum SNPWCs also smooth out into their own paste of temporal infinity and thoughtlessness.

Are we to infer that only finite portions of matter existing within standard eventuality can possess thought? Yes, because consciousness has no medium through vacuum. Faced with an infinite vacuum of macrocosm, into which are immersed an infinite quantity of finite eventualities, how can our finite consciousness form an accurate comprehension of the immensity? It would seem to suggest that our faculties, even at their most prodigious, may only skim the surface of reality. The disruption between these contours of time and eventuality – one expression without form or boundary and the other shaped and limited – is the greatest single contributor to quantum weirdness and entropy.

What rules can we extract from time to aid our struggling perception of it?

A particle or assembly of interconnected particles is time independent of un-associated particles or assemblies of particles.

In vacuum time does not exist in a measurable manner; it is infinite. Where vacuum symmetry has broken time becomes measurable as the rate of entropy, proportional to the quantity of eventuality present multiplied by its quantum deterioration.

Efficiency in minimising quantum deterioration can vary significantly between different arrangements of eventuality.

Each eventuality is unique, a specific temporal edifice with an unrepeatable sequence of queventa. It may be possible to duplicate an eventuality by precisely copying its framework of queventa, but this is only reproduction, not the original event itself.

Specific particles or particle assemblies can only occupy one temporal edifice between one isochron or isoflux and the next. When the edifice arrives at an isochron or isoflux that marks its termination all particles and particle assemblies within it must pass to other edifices according to CAM.

The creation of every temporal edifice depends on interactions between previous edifices. The quantum deterioration of each edifice depends on its internal reactions and external interactions, but it has no influence on its own previous configurations or the previous configurations of any other edifices.

A temporal edifice cannot enter into a new configuration whilst the edifices required to activate that configuration are substantially incompatible with it.

Mass distorts space-time by interfering with the pulse-wave natures of other mass in proximity.

Consciousness is the ability of an edifice to comprehend and influence its own interactions within the CSFC.

Intelligence is the motivation of consciousness to control its interactions with other edifices in order to reduce or eradicate negative interactions.

There are three stages to the activation of a temporal edifice as a living eventuality.

First life (existence). A temporal edifice's continued duration as an organised system, according to its intrinsic quantum deterioration, the value of its routine interactions with other edifices and the occurrence of non-routine external corruption, stabilisation or rejuvenation.

Second life (life as we experience it). A temporal edifice's continued duration as an organised system, resulting from its own work to combat quantum deterioration, stabilise and improve the value of its routine interactions with other edifices and influence the occurrence of non-routine external corruption, stabilisation or rejuvenation.

Third life (super-life). A temporal edifice's continued duration as an organised system, resulting from sustained self re-organisation to neutralise entropy without the need for routine or non-routine interactions with other edifices.

The Specific Problems of Time Travel

Although I have already considered the concept of time travel, and provided logical evidence to discount it as a viable possibility, I shall now explore the mathematics associated with hypothetical time travel, in order to illustrate the absurdity of the concept itself. To do this I will disregard logical proof and conceptual sanity, assuming instead that the cosmos is beyond all reasonable terms and evidence a hive of anarchy.

To a certain extent the equations of Einstein's Relativity, and hence relativistic effects such as travel at a significant proportion of light speed and hypothetical time travel, begin with the Lorentz equation and the Fitzgerald contraction.

Relativity contraction of length (Fitzgerald Contraction); $L_1 = L_0 \sqrt{1 - \dfrac{V^2}{C^2}}$

Relativity mass increase (Lorentz Equation); $M_1 = \dfrac{M_0}{\sqrt{1 - \dfrac{V^2}{C^2}}}$

Or $M_1 = M_0 \left(1 - \dfrac{V^2}{C^2}\right)^{-0.5}$

Using binomial theorem, $\left(1 - \dfrac{V^2}{C^2}\right)^{-0.5} = 1 + \dfrac{0.5V^2}{C^2}$

Substituting this in the Lorentz equation; $M_1 = M_0 \left(1 + \dfrac{0.5V^2}{C^2}\right) = M_0 + \dfrac{0.5MV^2}{C^2}$

Since $0.5MV^2 = E$; $M_1 = M_0 + \dfrac{E}{C^2}$ and $M_1 - M_0 = \dfrac{E}{C^2}$, therefore $M = \dfrac{E}{C^2}$ and;

Finally we arrive at $E = MC^2$

Considering Time Dilation in Time Travel -

Imagine a return voyage to the vicinity of the Andromeda Galaxy, at nearly the speed of light, during which the astronaut ages 26 years; using the relativity formula $N = \dfrac{R}{a}$ where N = non-relativistic time,

R = relativistic time and $a = \sqrt{1 - \dfrac{V^2}{C^2}}$ $\therefore N = \dfrac{R}{\sqrt{1 - \dfrac{V^2}{C^2}}}$

For the round trip to just outside the Andromeda Galaxy N = 4,000,000 (years light takes to travel from Earth to Andromeda and back again) and R = 26.

Since N = $\dfrac{R}{\sqrt{1-\dfrac{V^2}{C^2}}}$ it follows that 4,000,000 = $\dfrac{26}{a}$ and a = $\dfrac{26}{4,000,000}$ = 0.0000065

Thus $\sqrt{1-\dfrac{V^2}{C^2}}$ = 0.0000065 (v = proportion of c expressed as a unit value of 1)

1 - $\dfrac{V^2}{C^2}$ = 0.0000065^2 = 4.225 x 10^{-11}

$\dfrac{V^2}{C^2}$ = 1 - (4.225 x 10^{-11}) = 0.99999999995775

∴ V = $\sqrt{0.99999999995775}$ = 0.999999999978872 times the speed of light; this is the velocity at which the journey must be travelled in order that the astronaut only ages 26 years. Since non-relativistic observers, such as those left on Earth, see the astronaut's round trip taking four million years whilst he only ages 26 years, they will conclude correctly that the time dilation is 0.0000065 astronaut years (205.1277696 seconds)to each Earth year. Or 153846.1538 Earth years to each astronaut year. It seems as though the astronaut has leapt forward in time; his space journey has successfully disconnected him temporarily from interaction with the universe. His own biological clock has slowed down significantly, but he has still perceived the passing of time. By travelling at relativistic speed the astronaut has caused a warp in his own existence at a factor of 0.0000065.

Time Dilation Gamma Factor.

$\dfrac{1s}{\sqrt{1-v^2}}$ (where v is velocity as ratio to 100% c and 1s is 1 second)

% of c	ratio	Gamma factor	reciprocal
0	0	1	1
1	0.01	1.000050005	0.999949997
10	0.1	1.005037815	0.994987437
20	0.2	1.020620726	0.979795897
30	0.3	1.048284837	0.953939201
40	0.4	1.091089451	0.916515139
50	0.5	1.154700539	0.866025403
60	0.6	1.25	0.8
70	0.7	1.400280086	0.714142841
80	0.8	1.66666666	0.6
90	0.9	2.294157341	0.435889893
95	0.95	3.202563086	0.312249899
99.9	0.999	22.36627245	0.044710177
99.999	0.99999	223.6067978	0.0044721359
99.99999	0.9999999	2236.067978	0.0004472135
99.9999999	0.999999999	22360.67978	0.0000447213
100	1	Infinity	zero

However, if instead of getting into a star ship and travelling to Andromeda and back at virtually the speed of light, the astronaut had climbed into a time machine and set the controls to send him 4,000,000 years into the future, the effect would have been the same. In his departure year he would disappear from existence as far as non-relativistic observers were concerned, and suddenly 4,000,000 years later he would reappear. In an extraordinary parallel he would have warped time and space to allow himself to exist detached from regular universal interactions. This detachment should carry the same mathematical relationship as relativistic space travel, since T, V and M are all relativistic transformations of eventuality. It is reasonable, therefore, to assume that the same relativity time dilation factor (tdf = 0.0000065) is involved.

Thus the same requirements of life span and duration are required; the physical journey to Andromeda would not have been successful if our intrepid astronaut had set out ten years before he was due to expire. Similarly a time travel journey across 4,000,000 years is not practical if it is undertaken ten years before the astronaut will die, because the actual relativistic chasm of 26 time-dilated years must still be crossed. There is no point in the time machine popping into existence 4,000,000 years in the future with a '16 years dead' body inside.

To derive suitable equations that will describe the geometry of time travel, we need either direct or derived values for the following factors;

H = temporal shift (destination time less source time)
H_1 = dilated time that astronaut ages during temporal shift (tdf x H)
L = life span, A = age at displacement, A_1 = age at re-emergence
D_1 = original termination date, D_2 = revised termination date
T_1 = displacement date, T_2 = re-emergence date
P_1 = % life at displacement, P_2 = % life at re-emergence, P_3 = % in transition
S = source date (birth)

We can now derive the mathematical relationships necessary to proceed.

$$H = T_2 - T_1 = \frac{H_1}{tdf}$$

$$D_1 = S + L = (L - A) + T_1$$

$$D_2 = (T_2 - H_1) + (L - A) = T_2 + L - (H_1 + A)$$

$$H_1 = \sqrt{1 - \frac{V^2}{C^2}} \times H$$

$$L - A = D_1 - T_1 = D_2 - S$$

$$T_1 = T_2 - H = D_1 - (L - A) = S + A$$

$$T_2 = T_1 + H = D_2 - (L - A)$$

$$P_1 = 100\frac{A}{L}$$

$$P_2 = P_1 + P_3 = 100\frac{A + H_1}{L}$$

$$P_3 = 100\frac{H_1}{L}$$

With our example of a four million year leap into the future, let us first consider a temporal adventurer who was born in 2000, is due to live to the age of 70 and embarks on his epic journey in the year 2030. We obtain the following;

H = 4,000,000, H_1 = 26 (tdf 0.0000065 x 4,000,000)

L = 70, A = 30, A_1 = 56

D_1 = 2070, D_2 = 4,002,044

T_1 = 2030, T_2 = 4,002,030

P_1 = 42.85714286, P_2 = 80, P_3 = 37.14285714

These results are straightforward, but if our temporal adventurer is to arrive in the future whilst he is still 26 the exact value of A, and thus T_1, is no longer known; $A_1 \neq H_1$ and $\dfrac{A}{S + A}$ = tdf, and our calculations proceed quite differently.

STEP	BASE EQUATION	APPLICATION
1	$\dfrac{A}{L} \times \dfrac{H}{S + A} = \dfrac{H_1}{L}$	$\dfrac{A}{70} \times \dfrac{4,000,000}{2000 + A} = \dfrac{26}{70} = 0.371428557$
2	$A \times \dfrac{H}{S + A} = H_1$	$A \times \dfrac{4,000,000}{2000 + A} = 26$
3	$A = \dfrac{H_1(S + A)}{H}$	$A = \dfrac{26(2000 + A)}{4,000,000}$
4	$A = \dfrac{H_1}{H \div (S + A)}$	$A = \dfrac{26}{4,000,000 \div (2000 + A)}$
5	$HA = H_1(S + A)$	$4,000,000 \times A = 26(2000 + A)$
6	$\dfrac{HA}{H_1} = S + A = T_1$	$\dfrac{4,000.000A}{26} = 2000 + A = T_1$
7	$A\left(\dfrac{H}{H_1}\right) = S + A = T_1$	$A\left(\dfrac{4,000,000}{26}\right) = 2000 + A = T_1$
8	$A = \dfrac{S + A}{H \div H_1}$	$A = \dfrac{2000 + A}{4,000,000 \div 26}$
9	$S = A\{(H \div H_1) - 1\}$	$2000 = A\{(4,000,000 \div 26) - 1\}$
10	$A = \dfrac{S}{(H \div H_1) - 1}$	$A = \dfrac{2000}{(4,000,000 \div 26) - 1} = 0.013000084$

If the time traveller had been born in 1990 the value of A would have been 0.012935084 and the difference between the two results; 0.000065, exactly ten times the time dilation factor, thus reflecting the ten-year difference between 1990 and 2000. Similarly if the source date had been 2013 instead of 2000 a result of 0.013084585 yielding a difference of 0.000084501 (thirteen times the time dilation factor) would have been obtained. The startling conclusion is that the displacement time has a significant effect on the journey; time is specific.

We can summarise the difference in required leaving ages between one displacement time and another as d $\sqrt{1-\dfrac{V^2}{C^2}}$ where d = difference in years.

THIS SPECIFICITY IS THE GREATEST SINGLE PROOF THAT TIME IS UNIDIRECTIONAL AND QUANTISED. FROM THE VERY EQUATIONS NECESSARY TO CALCULATE TIME TRAVEL VOYAGES SPRINGS THE MOST POWERFUL EVIDENCE OF THE UNIVERSAL TABOO ON TIME TRAVEL.

Since the progression of time precisely follows specific, uni-directional increments in quanta defined by its own dilation at light-proportional velocity, it is not possible for any matter to experience time shifts as a result of time travel. The perpetual increment advance of time, in relation to its own light-proportional velocity dilation, means that any specified co-ordinate in time once spent can never be regained. Similarly, any specified co-ordinate in time not yet spent cannot be previewed. This is The Universal Law of Temporal Specification.

Other questions posed by hypothetical time travel.

From the cosmic point of view what happens to mass and volume of the time traveller and his time machine?

At T_1 any time travelling mass would cease to exist, its mass, volume and eventuality all removed from the CSFC in violation of the conservation laws.

Its non-existence would continue for H time measurements.

At T_2, after H time measurements of universal oblivion, the time travelling mass would suddenly re-exist, presumably in co-existence with the contemporary configuration of its own particles.

If leaping into the past, however, normal existence at T_1 would terminate forever; at T_2 any time traveller would materialise miraculously in co-existence with the historic configuration of their own particles at that time – a completely duplicated set. This dual existence would persist for H time units whereupon the contemporary configuration would disappear, leaving the time-shifted version to continue its ageing at T_1, but H time units older.

From the traveller's point of view at T_1 the universe would cease to exist, until T_2 when it re-materialises. If T_2 is in the future then the traveller will have aged in accordance with time dilation theory, but if T_2 is in the past the traveller will have become younger. The ageing or rejuvenation is proportional to natural life span, displacement time, temporal shift and time dilation.

The impossibility of time travel is clearly shown by these conflicting circumstances; there is no logical way to explain them and if we take the next step forward an even greater degree of the bizarre is encountered. With travel faster than the speed of light or travel backwards in time how much more impossible can we get?
Imagine our astronaut travelling to Andromeda at 100 x light speed. What happens? For a start H = 40,000 and we would expect that H_1 = 0.286 but that is not what happens.

Our dilation equation $\dfrac{1s}{\sqrt{1-v^2}}$ gives a surprising result of ± 0.0100005, meaning that the astronaut is rejuvenated by ± 0.0100005 x 40,000 = 400.02 years. For twice light speed the answer is ± 0.577350269 and at only 1.1 x light speed ± 2.182178905, so that the greater the multiple of c the less the effect. It seems that even on the far side of the light barrier the most noticeable transformations are nearer to c. Of more concern, though, is the floating negative value for the gamma factor.

Since v^2 is now always greater than 1 we are square rooting a negative number and the result is indeterminate; we can obtain a numerical value but one side is positive and one negative and this ambiguity persists through the equation. It can mean one of two possibilities; if the result is positive the astronaut could age more than the universe, if negative he gets younger whilst the universe continues to age!

% of c	ratio	Gamma factor	reciprocal
110	1.1	± 2.182178905	± 0.458257568
120	1.2	± 1.507556723	± 0.663324958
130	1.3	± 1.203858531	± 0.830662386
140	1.4	± 1.020620726	± 0.979795897
150	1.5	± 0.89442719	± 1.118033989
160	1.6	± 0.800640768	± 1.2489996
170	1.7	± 0.727392967	± 1.374772708
180	1.8	± 0.668153104	± 1.496662955
190	1.9	± 0.61898446	± 1.615549442
200	2	± 0.577350269	± 1.732050808
250	2.5	± 0.43643578	± 2.291287847
300	3	± 0.35355339	± 2.828427125
400	4	± 0.258198889	± 3.872983346
500	5	± 0.204124145	± 4.898979486
1000	10	± 0.100503781	± 9.949874371
10000	100	± 0.0100005	± 99.99499988
100000	1000	± 0.0010000005	± 999.9995

But which is true? Since the effect is time dilation, evidence would tend to suggest that we are dealing with a negative value, and thus travel beyond the speed of light causes the traveller to become younger. Similarly, therefore, time travel into the past would cause rejuvenation by the same tdf as indicated above. Let us assume that our hypothetical time traveller is going 4,000,000 years into the past and we are limiting the energy of his displacement so that he only gets 26 years younger. Birth date is still 2000 but now he must leave at a time that will allow rejuvenation; if we want him to arrive in the past as a 26 year-old, displacement must be in 2052. He can only afford to use energies a minute fraction in excess of light speed.

$H = (-4,000,000)$, $H_1 = (-26)$ {tdf $0.0000065 \times (-4,000,000)$}

$L = 70$, $A = 52$, $A_1 = 26$

$D_1 = 2070$, $D_2 = (-3,997,904)$

$T_1 = 2052$, $T_2 = (-3,997,948)$

$P_1 = 74.2857142$, $P_2 = 37.1428571$, $P_3 = (-37.14285714)$

To obtain the necessary energies for spatial or temporal travel at 'near light' speed would be next to impossible and require resources on a galactic scale. To obtain the energies needed for full light speed would be out of the question because the entire macrocosm would be necessary. Beyond light speed needs more energy than is actually available from infinity. The new table for travelling into the past is:

STEP	BASE EQUATION	APPLICATION
1	$-0.5A = H_1 = -A_1$	$-0.5 \times 52 = -26$
2	$-0.5\dfrac{A}{L} = \dfrac{H_1}{L} = \dfrac{-A_1}{L}$	$-0.5\dfrac{52}{70} = \dfrac{-26}{70} = -0.371428557$
2a	$S + A = T_1$	$T_1 = 2052$, now the source date
2b	$A_3 = T_1 \times tdf$	$A_3 = 2052 \times 0.0000065$ (adjustment)
2c	$A_3 \times \dfrac{H}{T_1} = H_1$	$A_3 \times \dfrac{-4,000,000}{2052} = -26$
3	$A_3 = \dfrac{H_1 T_1}{H}$	$A_3 = \dfrac{-26 \times 2052}{-4,000,000}$
4	$A_3 = \dfrac{H_1}{H \div T_1}$	$A_3 = \dfrac{-26}{-4,000,000 \div 2052}$
5	$HA_3 = H_1 T_1$	$-4,000,000 \times A_3 = -26 \times 2052$
6	$\dfrac{HA_3}{H_1} = S + A = T_1$	$\dfrac{-4,000.000 A_3}{-26} = 2000 + A = T_1$
7	$A_3\left(\dfrac{H}{H_1}\right) = S + A = T_1$	$A_3\left(\dfrac{-4,000,000}{-26}\right) = 2000 + A = T_1$
8	$A_3 = \dfrac{T_1}{H \div H_1}$	$A_3 = \dfrac{2052}{-4,000,000 \div -26}$
9	$T_1 = A_3\left\{(H \div H_1) - 1\right\}$	$2052 = A_3\left\{(-4,000,000 \div -26) - 1\right\}$
10	$A_3 = \dfrac{T_1}{(H \div H_1) - 1}$	$A_3 = \dfrac{2052}{(-4,000,000 \div -26) - 1} = 0.013338086$

Since $\dfrac{1}{H \div H_1} = \dfrac{H_1}{H} = tdf$ and $\dfrac{1}{(H \div H_1) - 1} = tdf\,(1 + tdf)$

$(H \div H_1) - 1 = \dfrac{1}{tdf\,(1 + tdf)}$

The parallels between sub-light space travel and forward time travel (on the one hand) or ultra-light travel and backward time travel (on the other) are quite amazing and of considerable significance. In order to be viable within human life expectancy greater energies are required as larger journeys are undertaken (both temporal and spatial). This holds true for both sub-light and ultra-light shifts across space or time, with the greater energies and dilation effects occurring nearest of all to the speed of light. Paradoxically it is the removal of energy at ultra-light levels that results in more ageing for the traveller and higher velocity. This is the realm of the hypothetical tachyon. At exactly the speed of light time dilation is infinite and one could cross the macrocosm instantaneously.

Of course we know that there are several problems with the reaching and breaching of light speed. To reach c requires infinite energy; the entire infinite mass of the macro-cosmos. Infinite mass and zero volume (Lorentz equation and Fitzgerald contraction) paired with infinite time sounds a lot like the Fourth Rule of Existential Proof. It is also like conditions at the time of the Big Crunch/Big Bang, between the **Tenth Imperative Isochron** at A4: T - 10^{-23} seconds and the **First Consequent Isochron** at B1: T + 10^{-12} seconds. Indeed, anything involving relativistic transformations will obey a common set of determinants. Compare the relativistic transformation formula from scenario 1 in

section A; $F = \dfrac{1}{1 \pm p/100}$

How similar it is to our time dilation formula $T = \dfrac{1s}{\sqrt{1 - v^2}}$

And from scenario 2 in section A;

$a_2 = \dfrac{a_1}{bc_2 / bc_1}$ and $a = \left(1 - \dfrac{bc_1}{bc_2}\right)$ which are similar to

$A = \dfrac{S}{(H \div H_1) - 1}$ and, $1 - \dfrac{V^2}{C^2}$

Whilst in scenario 3 of section A; in cases where $b_2 = b_1 \sqrt{\dfrac{c_2}{c_1}}$

divide both sides by b_1 so $\dfrac{b_2}{b_1} = \sqrt{\dfrac{c_2}{c_1}}$ and square both sides so $\left(\dfrac{b_2}{b_1}\right)^2 = \dfrac{c_2}{c_1}$

then finally multiply both sides by c_1 so that $c_2 = \left(\dfrac{c_1 b_2}{b_1}\right)^2$

This resembles the following time travel equation sequence:

$A = \dfrac{H_1(S + A)}{H}$ (forward time travel) or $A_3 = \dfrac{H_1 T_1}{H}$ (backward time travel)

Square both sides; $A^2 = \left(\dfrac{H_1(S + A)}{H}\right)^2$ or $A_3{}^2 = \left(\dfrac{H_1 T_1}{H}\right)^2$

Then divide both sides by (S + A) or T_1 so $\dfrac{A^2}{S + A} = \left(\dfrac{H_1}{H}\right)^2$ or $\dfrac{A_3{}^2}{T_1} = \left(\dfrac{H_1}{H}\right)^2$

Square root both sides: $\sqrt{\dfrac{A^2}{S + A}} = \dfrac{H_1}{H}$ or $\sqrt{\dfrac{A_3{}^2}{T_1}} = \dfrac{H_1}{H}$

Finally multiply both sides by H so: $H_1 = H\sqrt{\dfrac{A^2}{S + A}}$ or $H_1 = H\sqrt{\dfrac{A_3{}^2}{T_1}}$

Light, Mirror and Earth

We have seen how strange things happen when light speed is approached, reached or even exceeded. These flights of fancy provide us with interesting insights into the way the universe could be if ungoverned by natural logic and lawfulness. Of course, to achieve such bizarre effects would require prodigious quantities of energy, approximately equivalent to MAP for the simplest effects, but far in excess of that for more spectacular results. The speed of light not only restricts the movement of information and matter through space, it also restricts their movement through time. Time travel, as we have proven to ourselves so many times, is no more than the bright idea of science fiction.

Relativity, however, does have practical outcomes in reality, in connection with light, mass, space and time. Relativistic effects take hold near substantial gravity wells. And then there is the light, mirror and Earth experiment. When a mirror and Earth are travelling in the same direction at the same speed and in the same plane, the time taken by the beam of light to journey to the mirror and back (which is in the direction of Earth's travel) is:

$$\frac{d}{c+v} + \frac{d}{c-v} = \frac{2dc}{c^2-v^2}$$ where d = distance, c = speed of light, v = velocity of

Earth & mirror.

If the mirror is travelling at right angles to Earth; $\frac{y}{x} = \frac{v}{c}$ therefore $y = \frac{vx}{c}$

M   y   M1

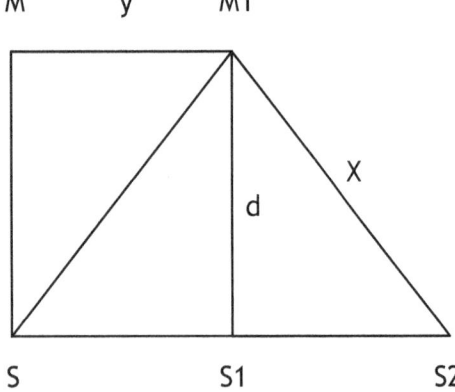

S   S1   S2

To solve the value for 'x' use Pythagoras's theorem;

$$x^2 = d^2 + \frac{vx^2}{c} \quad \text{so } x^2 - \frac{vx^2}{c} = d^2$$

$$x^2 - \frac{v^2x^2}{c^2} = d^2 \quad \text{then} \quad \frac{c^2x^2 - v^2x^2}{c^2} = d^2 \quad \text{and } (c^2-v^2)x^2 = d^2c^2$$

therefore $x^2 = \frac{d^2c^2}{c^2-v^2}$ and by square rooting each term $x = \frac{dc}{\sqrt{c^2-v^2}}$

The total path travelled by the light is 2x or $\dfrac{2dc}{\sqrt{c^2-v^2}}$ and the time taken for the

light to cover this distance is $\dfrac{2dc}{\sqrt{c^2-v^2}} \div c = \dfrac{2d}{\sqrt{c^2-v^2}}$

Now let us compare the parallel and perpendicular cases: $\dfrac{2dc}{c^2-v^2}$ and $\dfrac{2d}{\sqrt{c^2-v^2}}$

$$\frac{2dc}{c^2-v^2} \div \frac{2d}{\sqrt{c^2-v^2}} = \frac{2dc}{c^2-v^2} \times \frac{\sqrt{c^2-v^2}}{2d} = \frac{c\sqrt{c^2-v^2}}{c^2-v^2}$$

a square root divided by the number (a) it is the square root of is $\dfrac{1}{\sqrt{a}}$

so with $\dfrac{c}{\sqrt{c^2-v^2}}$ and $\dfrac{c\sqrt{1 \div c^2}}{\sqrt{c^2-v^2}\sqrt{\dfrac{1}{c^2}}} = \dfrac{c \div c}{\sqrt{(c^2 \div c^2)-(v^2 \div c^2)}} = \dfrac{1}{\sqrt{1-(v^2 \div c^2)}}$

This appears to be remarkably closely related to the relativistic time dilation factor equation.

Particle Families

Particles are divided into groups or families, according to their relatedness and properties.

QUARKS and LEPTONS are the basic building blocks of matter; elementary units from which all matter is made. They have a spin of ½ (leptons have left-handed spin). There are six quarks and six leptons, and the quarks combine to make baryons (3 quarks) and

mesons (2 quarks) and the recently discovered penta-quark (uudd $\bar{S}$). Quarks undergo transformations by exchanging W bosons, thus causing hadrons to decay by means of the weak force. Up and down quarks are the least massive but most stable, and have therefore become the main mass constituents of the universe. Quark decay is always from one with charge + ⅔ to one with charge - ⅓ or vice versa.

BOSONS are intermediary force carrying particles.

HADRONS are particles that are divisible into quarks and are grouped into sub-families of 1, 8, 10 and 27 members, all members of a specific family having identical spin but different electric, isotopic and strange charges. The two main sub-divisions of the hadrons are;

BARYONS, with spins of $\dfrac{1}{2}$ or $\dfrac{3}{2}$ and constructed from 3 quarks, mostly decay by the

strong force interaction at very short time scales (10^{-23} seconds).

MESONS, with spins of 0, 1 or 2 are constructed from 1 quark and 1 anti-quark.

Elementary Particles: THE QUARK FAMILY

Name	Symb	Mass	Life (s)	B No	Charge	Spin	S	C	B	T
Up	U	360 MeV		⅓	+ ⅔	½	0	0	0	0
Down	D	360 MeV	900	⅓	- ⅓	½	0	0	0	0
Strange	S	540 MeV	1.24×10^{-8}	⅓	- ⅓	½	-1	0	0	0
Charm	C	1500 MeV	1.1×10^{-12}	⅓	+ ⅔	½	0	1	0	0
Bottom	B	5 GeV	1.3×10^{-12}	⅓	- ⅓	½	0	0	1	0
Top	T	174 GeV		⅓	+ ⅔	½	0	0	0	1

Elementary Particles: THE LEPTON FAMILY

Name	Symb	Anti-p	Mass MeV/c^2	Lepton No	Chg	Life (sec)
Electron	e^-	e^+	0.511	1	-1	stable
e-neutrino	v_e	$\bar{v}_e$	$<7 \times 10^{-6}$	1	0	stable
Muon	μ^-	μ^+	105.7	1	-1	2.2×10^{-6}
Mu-neutrino	v_μ	$\bar{v}_\mu$	<0.27	1	0	stable
Tau	τ^-	τ^+	1777	1	-1	2.96×10^{-13}
Tau-neutrino	v_τ	$\bar{v}_\tau$	<31	1	0	stable

Elementary Particles: Overview with Bosons

Generation	Leptons (main)	Leptons (sub)	Quarks - ⅓	Quarks + ⅔
I	e	ν_e	D	U
II	μ	ν_μ	S	C
III	τ	ν_τ	B	T
Force carrying Bosons	W^+, W^- (weak)	Z^0 (weak)	Gluon (g) (strong)	Photon (γ) (el-mag + light)

THE HADRON - BARYON (Fermion) FAMILY

Name	Sym chg	Qs	Mass MeV/c^2	spin	B No	S chg	Life (s)	decay
Proton	p +1	uud	938.3	½	1	0	stable	
Neutron	n 0	ddu	939.6	½	1	0	920	$pe^-\bar{\nu}_e$
Lambda	Λ^0	uds	1115.6	½	1	-1	2.6×10^{-10}	$p\pi^-$, $n\pi^0$
Sigma	Σ^+	uus	1189.4	½	1	-1	0.8×10^{-10}	$p\pi^0$, $n\pi^+$
Sigma	Σ^0	uds	1192.5	½	1	-1	6×10^{-20}	$\Lambda^0\gamma$
Sigma	Σ^-	dds	1197.3	½	1	-1	1.5×10^{-10}	$n\pi^-$
Delta	Δ^{++}	uuu	1232	1.5	1	0	0.6×10^{-23}	p, π^+
Delta	Δ^+	uud	1232	1.5	1	0	0.6×10^{-23}	p, π^0
Delta	Δ^0	udd	1232	1.5	1	0	0.6×10^{-23}	n, π^0
Delta	Δ^-	ddd	1232	1.5	1	0	0.6×10^{-23}	n, π^-
Xi cascade	Ξ^0	uss	1315	½	1	-2	2.9×10^{-10}	$\Lambda^0\pi^0$
Xi cascade	Ξ^-	dss	1321	½	1	-2	1.64×10^{-10}	$\Lambda^0\pi^-$
Omega	Ω^-	sss	1672	1.5	1	-3	0.82×10^{-10}	$\Xi^0\pi^-$, $\Lambda^0 K^-$
Lambda	Λ_C^+	udc	2281	½	1	0	2×10^{-13}	

THE HADRON - MESON (Boson) FAMILY

NAME	SYM CHG	ANTI -P	QS	MASS MEV/C^2	S	C	B	LIFE (S)	DECAY
Pion	π^+	π^-	$u\bar{d}$	139.6	0	0	0	2.6×10^{-8}	$\mu^+ \nu_\mu$
Pion	π^0	self	$\dfrac{u\bar{u}+d\bar{d}}{\sqrt{2}}$	135	0	0	0	0.83×10^{-16}	2γ
Kaon	K$^+$	K$^-$	$u\bar{s}$	493.7	1	0	0	1.24×10^{-8}	$\mu^+ \nu_\mu$, $\pi^+ \pi^0$
Kaon	K^0_S	K^0_S	$d\bar{s},\ \bar{d}s$	497.7	1	0	0	0.89×10^{-10}	$\pi^+ \pi^-$, $2\pi^0 \pi^0$
Kaon	K^0_L	K^0_L	$d\bar{s},\ \bar{d}s$	497.7	1	0	0	5.2×10^{-8}	$\pi^+ e^- \bar{\nu}_e$
Eta	η^0	self	$(u\bar{u}+d\bar{d} - 2s\bar{s})/\sqrt{6}$	548.8	0	0	0	$<10^{-18}$	$2\gamma, 3\mu$
Eta prime	$\eta^{0'}$	self	$(u\bar{u}+d\bar{d} - 2s\bar{s})/\sqrt{6}$	958	0	0	0		
Rho	ρ^+	ρ^-	$u\bar{d}$	770	0	0	0	0.4×10^{-23}	π,π
Rho	ρ^0	self	$u\bar{u},\ d\bar{d}$	770	0	0	0		
Omega	ω^0	self	$u\bar{u},\ d\bar{d}$	782	0	0	0		
Phi	Φ	self	$s\bar{s}$	1020	0	0	0	20×10^{-23}	$K^+ K^-$, $2K^0$
D	D$^+$	D$^-$	$c\bar{d}$	1869.4	0	1	0	10.6×10^{-13}	K^+-, e^+-
D	D^0	$\bar{D}^0$	$c\bar{u}$	1864.6	0	1	0	4.2×10^{-13}	$(K, \mu, e)+$
D	D^+_S	D^-_S	$c\bar{s}$	1969	1	1	0	4.7×10^{-13}	K^+-
J/Psi	J/Ψ	self	$c\bar{c}$	3096.9	0	0	0	0.8×10^{-20}	$e^+ e^-$, $\mu^+ \mu^-,...$
B	B$^-$	B$^+$	$b\bar{u}$	5279	0	0	-1	1.5×10^{-12}	D^0 +-
B	B^0	$\bar{B}^0$	$d\bar{b}$	5279	0	0	-1	1.5×10^{-12}	D^0 +-
B$_S$	B^0_S	B^0_{-S}	$s\bar{b}$	5375	0	0	-1	1.5×10^{-12}	
Upsilon	Y	self	$b\bar{b}$	9460.4	0	0	0	1.3×10^{-20}	$e^+ e^-$, $\mu^+ \mu^-,...$

Useful Measurements

PLANCK and BOHR values

Planck volume = 10^{-99} cm^3

1 Googol Planck volume = 10cm^3

Planck density = 10^{94} gm/cm^3 (1 Planck mass occupying a volume 1 Planck length in diameter)

Planck length = 10^{-33} cm (10^{-20} x proton size) (1Å = 10^{-8} cm)

Planck mass = mass of particle with wavelength of 1 Planck length = 10^{-5} gm (5.978715772 x 10^{19} x proton mass)

Planck energy (1000kwh) = energy needed to probe distances as small as Planck length

Planck tension = 10^{39} tons

Plank time = time light would take to cross a distance of 1 Planck length = 10^{-43} seconds = time at which universe was 1 Planck length in diameter

Planck constant = $\hbar$ = 6.6262 x 10^{-27} erg seconds = 6.6262 x 10^{-34} joule seconds = 1.05 x 10^{-27} gmcm/s

1 Planck Electron Neutrino Event (PENE) = 1.247884932 x 10^{-174} gmcm^3 s

Bohr radius = 0.52917 Ångstrom = radius of 1st electron shell (Hydrogen atom)

Bohr volume = 0.6206892934 Å^3 = 6.206892934 x 10^{-25} cm^3 (Hydrogen atom)

Bohr volume can contain 4.032398976 x 10^{37} protons

Bohr density when thus filled is 1.086629107 x 10^{38} gm/cm^3

Atomic mass unit (AMU) = 1.66 x 10^{-24} gm = $\dfrac{1}{12}$ mass of carbon atom.

ELECTRON SHELL COMPLETION

Shell	n	Sub-orbital s=2, p=6, d=10, f=14	Max e 2n^2	Aufbau filling order
K	1	1s	2	1s
L	2	2s, 2p	8	2s, 2p
M	3	3s, 3p, 3d	18	3s, 3p, 4s
N	4	4s, 4p, 4d, 4f	32	3d, 4p, 5s
O	5	5s, 5p, 5d, 5f, 6s, 6p, 6d	50	4d, 5p, 6s, 4f, 5d, 6p, 7s, 5f, 6d, 7p
Oa	2	7s, 7p	8	

Low frequency / longer wavelengths disturb particle velocity less but cannot resolve location so well. Shorter / high frequency wavelengths determine position better but interfere with velocity. Uncertainty values of position and velocity are always higher than Planck's constant.

GRAVITY EQUATIONS

A = combined surface areas of electron shells, in electron area units.
Z = 1 electron area unit (EAU), E = number of electrons present,
O = number of spherical orbits transited per electron per second

Ip = $\dfrac{A}{Z}$ = No of orbital insertion points per electron per orbit

IpO = total number of insertion points per electron per second

Sei (Single Electron Incident) = 1/IpO = time each electron spends at an OIp
Tei (Total Electron Incident) = time all electrons spend at an OIp = E/IpO

1/O = duration of orbit

Portion of nucleus exposed by electron shells at any one time: U = $\dfrac{A - ZE}{A}$

P = number of protons in nucleus (= E)
Fpc = Force Proton Charge (per atom) = {P - (E/Ip)} x U

Fer = Force Electron Resistance = P - Fpc

Gravitational force between two objects; F = $\dfrac{Gm_1 m_2}{r^2}$

G = $\dfrac{Fr^2}{m_1 m_2}$

Acceleration due to gravity; F = Mg where g = $\dfrac{Gm}{r^2}$

Equivalence of centrifuge and gravity in orbital situations; $\dfrac{M_1 v^2}{r} = \dfrac{GM_1 m_2}{r^2}$

Orbital period of a planet in relation to axis of elliptical orbit;

$$T^2 = \left(\dfrac{4\pi^2}{GM}\right) r^3$$

Escape velocity; V = $\sqrt{\dfrac{2GM}{R}}$

Velocity equation: $\dfrac{dv}{dt} = g$, t = $\dfrac{v}{g}$, v = gt

Distance equation: v = $\dfrac{dx}{dt}$ = gt , x = $\dfrac{gt^2}{2}$

Time equation: $t = \sqrt{\dfrac{2x}{g}}$

Firing an object up against gravity: $x = \dfrac{v^2}{2g}$

Potential energy between two particles: $U = \dfrac{GM_1 m_2}{r}$

Total energy of system = kinetic energy + potential energy

G (gravitational constant)= 6.673×10^{-11} N (m^2/kg^2)= 6.673×10^{-8} dyne(cm^2/gm^2)
M,m = masses of objects involved
r = distance between objects

Gravity acting on a mass m_1 that is Ø times further away than another mass m_2 does so $Ø^2$ times as weakly as it does on m_2.

Equation for velocity of objects rolling down an inclined plane:

V = t(g sin A) + u, thus $g = \dfrac{v}{t \sin A}$ indicating constant acceleration

Distance an object falls in a given time is thus d = ½gt^2 or d = $16t^2$

Where t = time, A = angle of incline, u = initial velocity, g = $32ft/s^2$ or $9.82m/s^2$

NEWTON'S LAWS OF MOTION

1: Every object maintains a state of rest, or uniform motion in a straight line, unless compelled by external force to change that state. Objects at rest tend to remain so.

2: f = ma (in Kg m/sec^2 = 1N)

3: For every action there is an equal and opposite reaction.

BLACK HOLE MEASUREMENTS

Black hole entropy formula: $S = \dfrac{Akc^3}{4\hbar G}$

Black Hole Temperature: $T = \dfrac{\hbar c^3}{8\pi kGM}$

Black Hole radius: $R = \dfrac{2GM}{C^2}$

Black Hole density: $1.426903643 \times 10^{27}$ gm/cm^3

Earth black hole radius; 1cm
Sun black hole radius; 69.42495068 cm

A = area of event horizon, K = Boltzman's constant
ℏ = Planck's constant, c = speed of light, S = entropy

MISCELLANEOUS MEASUREMENTS

Cosine θ = Sine θ = $\dfrac{\sqrt{2}}{2}$ = 0.70710678

(Where θ is our 45° angle at both extremities of the graph).

The general wave equation is $X = A\cos(\dot{\omega} \times t + d/\lambda)$

Where; **X** = distance from equilibrium point, A = amplitude of motion
$\dot{\omega}$ = angular frequency of motion, t = time, d = spatial position on wave
λ = wavelength

Schrodinger equation: $i\hbar \dfrac{d}{dt} 4(\vec{x},t) = H4(\vec{x},t)$ $\therefore$ $i\hbar \dfrac{d}{dt} = H$

Hubble's law: V = Hd
Where V = velocity, H = Hubble's constant, d = distance from Earth

UNIVERSAL CONSTANTS

Planck constant = ℏ = 6.6262 x 10^{-27} erg seconds = 6.6262 x 10^{-34} joule seconds = 1.05 x 10^{-27} gmcm/s

Boltzman's constant = 5.67 x 10^{-8} Watts m^{-2} K^{-4} = 1.38 x 10^{-23} J/K
Hubble's constant = 75 km/s/Megaparsec

Cosmological constant (λ) = $\dfrac{8\pi G}{3c^2}\rho$ = 10^{-35} s^{-2}

Event Relativity Constant = $\sqrt[3]{Pe}$

Speed of Light = c = 2.997925 x 10^{10}
c^2 = 8.987554306 x 10^{20}
c^3 = 2.694401374 x 10^{31}

UACsi = $\dfrac{1}{c^3}$ = 3.711399533 x 10^{-32}

MID per Q = $\sqrt[3]{1/c}$

GEOMETRY

To calculate radius from volume: $R = \sqrt[3]{v \div \left(\frac{4}{3}\pi\right)}$

Area of sphere $= 4\pi r^2$

EINSTEIN PERMUTATIONS

$E = mc^2$

$E = \hbar v$

$mc^2 = \hbar v$

$v = \dfrac{mc^2}{\hbar}$

Time dilation Gamma factor: $\dfrac{s}{\sqrt{s-v^2}}$ where s = time, v = velocity as % of light speed

(ie; 10% = 0.1)

In a plane flying east a clock records less elapsed time than it would in a plane flying west. Also clocks at higher altitudes tick faster; nearer a centre of gravity they tick slower.

SOLAR & PLANETARY INFORMATION

Body	Mass (grams)	Density	gravity	Volume (cc)	From Sun
Sun	1.99×10^{33}	1.409		1.4123×10^{33}	Nil
Mercury	3.28735×10^{26}	5.44	0.38kg	6.4999×10^{25}	58×10^6 km
Venus	4.871255×10^{27}	5.25	0.88kg	0.9533×10^{27}	108×10^6 km
Earth	5.977×10^{27}	5.52	1kg	1.0833×10^{27}	150×10^6 km
Mars	6.45516×10^{26}	3.94	0.38kg	1.62498×10^{26}	228×10^6 km
Jupiter	1.9×10^{30}	1.333	2.64kg	1.42565×10^{30}	778×10^6 km
Saturn	5.6901×10^{29}	0.71	1.16kg	8.17907×10^{29}	1427×10^6 km
Uranus	8.72642×10^{28}	1.27	1.17kg	7.25825×10^{28}	2870×10^6 km
Neptune	1.02804×10^{29}	2.06	1.18kg	6.17493×10^{28}	4497×10^6 km
Pluto	1.3×10^{25}	2.03	0.06kg	1.68332×10^{26}	5900×10^6 km

GENERAL MEASUREMENTS

$\pi = 3.14159265358979323846$

Base $\log_e = 2.71828182845904523536028747$

Quadratic equations; for $ax^2 + bx + c = 0$, then $x = \dfrac{-b \pm \sqrt{b^2 - 4ac}}{2a}$

Electron mass = 9.10956 x 10^{-28} gm

Proton mass = 16726 x 10^{-28} gm

Proton radius = 1.5431473299 x 10^{-21} cm

Proton volume = 1.539255657 x 10^{-62} cm^3

Neutron mass = 16749 x 10^{-28} gm

Nuclear density = 2.44 x 10^{14} gm/cm^3

Radius of atomic nuclei = (1.3 x 10^{-13} cm) $\sqrt[3]{M}$

Avogadro's number = 6.0222 x 10^{23} /mol

PREFIXES

Tera	=	10^{12}
Giga	=	10^{9}
Mega	=	10^{6}
Kilo	=	10^{3}
Hecto	=	10^{2}
Deka	=	10^{1}
Deci	=	10^{-1}
Centi	=	10^{-2}
Milli	=	10^{-3}
Micro	=	10^{-6}
Nano	=	10^{-9}
Pico	=	10^{-12}
Femto	=	10^{-15}
Atto	=	10^{-18}

1 Ångstrom = 10^{-10} m

1 kilobar = 10^{8} N/m^2 = 1.45 x 10^{4} $lbft/in^2$

1 newton = $1kg/s^2$

1 pascal (pressure) = 1 N/m^2

1 dyne = 10 micronewtons

1 poise (viscosity) = 0.1 Ns/m^2

1 Hertz = 1cycle/s

1 atmosphere = 760 torr = 1.01325 x 10^5 Pa = 14.69 lbft/in^2

1 tesla (magnetic induction) = 10000 gauss = 1 webber/m^2

1 webber = 1 v/s

1 gamma = 10^{-9} tesla = 10^{-13} gauss

1 kelvin, 1^0K = 1^0C, 0^0C = 273.16^{0}K
1^0 rankine = 1^0 Fahrenheit

$$^0C = \frac{5}{9}(^0F - 32)$$

1 radian = plane angle with vertex at centre of circle, subtended by arc equal in length to radius

1 steradian = solid angle with vertex at centre of sphere, subtended by area of surface equal to that of a square with sides equal to radius.

1 joule = 1N/m = 1 volt coulomb

1 watt = 1j/s, 1kw = 1.341 Hp, 1kwh = 3.6 x 10^6j

1 Hp = 745,700 w, 1 BTU = 1055.06j

1 heatflow unit = 0.042 w/m^2 = 1 microcal/cm^2/s

1 calorie = heat required to raise temperature of a certain volume of water from 14.5^{0}C to 15.5^{0}C = 4.184 joules

1eV = 1.602 x 10^{-19}j
300 GHz (wavelength 1m) has energy of 0.0012 eV

1 volt = 1w/A = 1j/coulomb = 1j/Ampsec; 1 farad = 1As/V

1 ohm = 1v/A resistance; 1 henry = 1Vs/A; 1 coulomb = 1 Ampsec

1 Ampere = current which, if maintained in each of two parallel wires in free space, would produce a force between them of 2 x 10^{-7}N/m due to magnetic field = 1 coulomb/s.

1 mole = amount of substance in a system with as many elementary entities as atoms in 0.012 kg of carbon 12 = 6.022 x 10^{23} = Avogadro's number

Concentration = mol/m^3

1 candela = luminous intensity of black body of $\dfrac{1}{600000}$ m^2 at temperature of freezing platinum (2045^{0}K)

1 lumen (light flux) = a source radiating 1 candela in all directions, radiates a light flux of 4π lumens.

1 lux = 1 lumen/m^3

density = gm/cm^3

1 litre = 1000cm^3

1 tonne = 1000kg

1 second = 9192631770 cycles of radiation associated with a specified transition of the caesium 133 atom.

Velocity = m/s

Acceleration = m/s^2

PAPER TWO:
Applied Quantum Eventuality

Entrée

Paper Two is concerned with the interaction between consciousness and reality. It explores the principles of Quantum Eventuality in the Multiple Field Continuum (MF Continuum), which is the sum of the 'Ex-world' (all that is perceived beyond oneself) and the 'In-world' (perception of oneself).

All that has ever been, is or ever will be perceived can be expressed as a simple statement of the interface between Quantum Eventuality and consciousness; the interception, expansion into consciousness, conception and translation of reality information into language.

Units of measurement used in Paper Two;

Unit of Quantum Interception Arc = Quinarc
Unit of Quantum Expansion Arc = Quexar
Unit of Quantum Concept Arc = Quonsar
Unit of Quantum Language Arc = Qualar

Section E: Selective Wave Interception

Principles of Interception Eventuality

Throughout the progression of this book I have tried to distance the reader from any notion of universal separatism. Even the macrocosm with its endless pulses of collapsing and expanding photo-cosmoses, has been shown as an organic entity held in communion by universal factors. Apart from the obvious consistency of mathematical principles, the actual operating medium of the entire macrocosm is the CSFC. It is a single, infinite operating platform from which endless quantum events can manifest. At local level all particles and even vacuum itself are administered by Sub-Nucleonic Pulse-Wave Contours. Information is universally transmitted at all frequencies by micro-photons. Gravity, electromagnetism, the weak and strong forces and every Pulse-Wave Interaction is a result of micro-photon information exchange.

Even entropy and quantum weirdness obey strict rules set down in the Cosmic Accounting Mechanism. There is no other way for them to behave; it is built into the nature of the universe to behave logically and with balance. However, the history of human understanding has always shown a wide discrepancy between reality and what is understood of reality. Progress is sometimes made slowly in a particular area of comprehension, in other areas progress may be rapid. But always amongst conscious beings there are those individuals who do not know an equal fullness of comprehension. The sum of human awareness of, for example, mathematics is not represented within every single person. And how many humans could have ever written the fugues of JS Bach?

It is likely that consciousness has evolved in many parts of the macrocosm, and equally likely that it has an uneven link with the universal paste of information that spreads throughout the heavens. It is not difficult to understand that creatures of limited and small size will not be able to cope with an entire macrocosm of information; only the macrocosm can cope with a macrocosmic pool of knowledge. Also it is not difficult to understand that different units of consciousness will have capacity in different ways. No matter how big the mind or how accurate, it is a finite quantity trying to grapple with infinity. Obviously it will not succeed in anything but a partial grasp of the finite and the local.

Interception Eventuality is a measure of the proportion of events that are identified and recognised by consciousness. It will always be significantly less than Q - with the ratio decreasing on average as Q increases. In order to accurately measure Interception Eventuality it is necessary to isolate a temporal edifice (hypothetically), calculate its minimum and maximum quantum weirdness levels, and apply the following formula; Q_i^a

$$= \frac{Q^1 + q_0}{2} \times A^0 \text{ where;}$$

Q_i^a = Quantum Interception Arc or Quinarc

Q^1 = eventuality less minimum loss of clarity

q_0 = eventuality less maximum loss of clarity

A^0 = activation of observer in relation to temporal edifice

This formulation gives rise to the Quantum Interception Arc (QIA) value for the entire event, and simple application of the TVM ratio will show how each factor was impacted. These calculations demonstrate how the Quinarc increases with observer sensitivity or parity with the event. The less attuned consciousness is to a particular temporal edifice, event or aspect of reality, the greater the fissure between that consciousness and that reality. For example, if a mind has a considerable grasp of chemistry then its interception arc for chemistry is high, equilibrium and parity is achieved and thus accuracy and understanding in processing chemical information is efficient. The identity of mind and fact is said to be symmetrical. But if a mind has only a basic grasp of general science then its interception arc for chemistry is low, disparity is dominant and thus accuracy and understanding in processing chemical information is inefficient. The identity of mind and fact is said to be asymmetrical.

When I refer to an interception arc I am acknowledging the extent to which an event's profile is at a tangent to the mind observing it. In Paper One, section A, we looked briefly at the relationship between reality and consciousness with the help of some graphs;

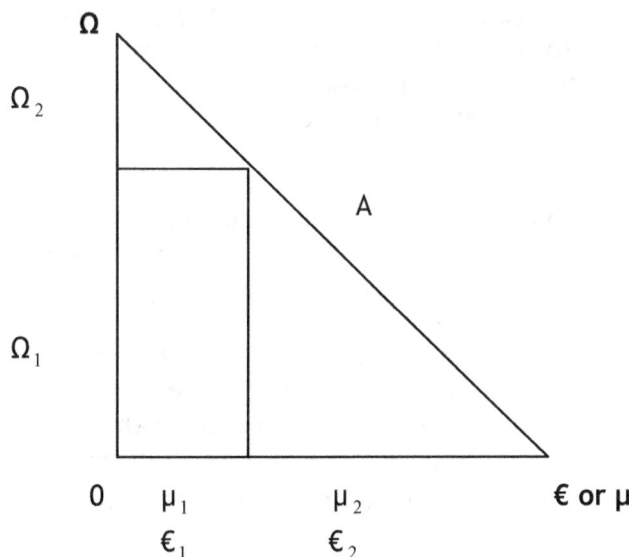

Where;
Ω = Total information in a system, Δ = Pattern strength
μ = Comprehension of each unit of information
$\emptyset$ = Information lost to Entropy per time unit
ϵ = Entropy, T = number of time units
Then $\Omega\mu$ = total comprehension and, $\Delta = \Omega\mu - \emptyset T$ and $\Omega\mu = \Delta + \emptyset T$

$\Omega = \dfrac{\Delta + \emptyset T}{\mu}$ and $\mu = \dfrac{\Delta + \emptyset T}{\Omega}$ so we can now expand this argument:

$\emptyset T = Q - \dfrac{Q^1 + q_0}{2}$ therefore $\Delta = \Omega\mu - \left(Q - \dfrac{Q^1 + q_0}{2}\right)$ with rearrangement of terms

$$\Delta = \Omega\mu + \frac{Q^1 + q_0}{2} - Q$$

These equations underline an important fact; the efficiency of a conscious mind's interception arc for a particular set of event-realities can only be determined by its natural affinity with that set and its strength of development for that affinity. If we see one interpretation of the Quantum Interception Arc as a function of identity, then the Interception Arc of the reality itself is 100% or unity (1) and is in fact the full identity of that reality. But when consciousness observes reality there is a loss in identity as reality is converted into perception.

When consciousness either lacks affinity with an event-reality or has failed to develop a latent affinity with it, there is a pronounced disparity between the consciousness and the reality it perceives. An example of this would be the situation of a mind untrained or ungifted in music encountering a symphony. There would be a clear mismatch between the consciousness and the reality, and it may even be said that - for all intents and purposes, as far as capacity to comprehend musical attributes is concerned - the consciousness is off-line. Thus a symphony is an extraordinary creation alongside an unmusical person, but quite usual alongside a composer – particularly the composer responsible for the symphony. The degree to which the Interception Arc of the symphony (and hence its identity) is perceived by the unmusical person now rests only on their latent affinity for symphonic music. If they have no such affinity then the mismatch is complete.

This latent affinity represents the person's potential to be stimulated by what they are witnessing, in this case a symphony. The composer has a significant stimulus potential and therefore exists on more equal terms with the identity of the symphony. The untrained or ungifted person has a low stimulus potential and will therefore struggle with symphonic comprehension in inverse proportion to their latent potential. Since Quantum Interception Arcs are comparative measures of the primary interaction between events and minds, it becomes clear that they are an expression of the initial degradation of information in its transition from the pure state to the corrupted. For this reason I shall call the product of Quantum Interception Arc 'view-eventuality'. O^{sp} = Observer's Stimulus Potential = $\sqrt{A^0}$.

$_vQ = Q_i^a \times O^{sp} \times Q = {_v}T {_v}V {_v}M$ (in same proportion as TVM)

Here is an example to illustrate how the Quantum Interception Arc represents the primary phase in entropy's degradation of information. Say;

Q^1 = 0.9
q_0 = 0.215
A^0 = 0.88
O^{sp} = 0.938083152

Then Q_i^a = $\dfrac{Q^1 + q_0}{2}$ x A^0 so Q_i^a = $\dfrac{0.9 + 0.215}{2}$ x 0.88 = 0.4906

$_vQ$ = Q_i^a x O^{sp} x Q = 0.4906 x 0.938083152 x Q = 0.460223594 x Q

Obviously we need to maximise all potentials if we are to avoid catastrophic degradation of reality within our perception, which is where education, logic, experience and ability come to our aid. Look how different the result would be if the factors were more efficient.

Q^1 = 0.997
q_0 = 0.339
A^0 = 0.999
O^{sp} = 0.999499874

$$Q_i^a = \frac{0.997 + 0.339}{2} \text{ x } 0.999 = 0.667332$$

$_vQ$ = 0.667332 x 0.999499874 x Q = 0.666998249 x Q

A significant improvement but even at these greater efficiencies degradation of 33% has occurred which is poor. One approach to correct this situation would be to improve the transmission or source of our information. This could be as simple as approaching a professor of mathematics for help in solving an equation, rather than asking a friend with no special ability in the subject. Next, we might replace the person perceiving the professor's explanation with an extremely gifted student of mathematics who is paying very particular attention to the explanation. Finally we might even replace the student listener with another professor of mathematics whose stimulus potential for the subject equation is phenomenal. The perfect scenario is for all the values to be one, an expression of totality.

Q^1 = 1
q_0 = 1
A^0 = 1
O^{sp} = 1

Now Q_i^a = 1 and $_vQ$ = Q

Of course this is an extremely unlikely ideal situation, even between two master professors of mathematics. Unfortunately entropy and quantum weirdness ensure that even at the highest level a quantum spanner is in the works.

Q^1 = 0.9999
q_0 = 0.9
A^0 = 0.9999
O^{sp} = 0.999949998

$$Q_i^a = \frac{0.9999 + 0.9}{2} \text{ x } 0.9999 = 0.949855005$$

$_vQ$ = 0.949807511 x Q

Far from being an exhortation to give up because there will always be a certain degradation of information, this is motivation to reach for the highest level in all dealings with reality. $_vQ$ = 0.949807511 x Q is preferable to $_vQ$ = 0.460223594 x Q to be sure. But it is exceedingly presumptuous to assume that 'eventuality less maximum loss of clarity' can ever be as high as 90% (0.9). With our new calculations of 'view-eventuality' based on a proportion of unity, Q is unity (1) and our other equations for the 'primary phase in entropy's conscious degradation of information' produce the following results for our first example.

$$ØT = Q - \frac{Q^1 + q_0}{2}, \text{ so } ØT = 1 - \frac{0.9 + 0.215}{2} = 0.4425$$

Since Q = 1 it follows that $\Omega = 1$ and so $\dfrac{\Delta + \phi T}{\mu}$ = 1 and $\mu = \Delta + \phi T$

We know that ØT = 0.4425 so $\Delta = \mu - \phi T$ and with rearrangement of terms

$$\Delta = \Omega \mu + \frac{Q^1 + q_0}{2} - Q \text{ so that } \Delta = \mu + 0.5575 - 1 \text{ and } \Delta = \mu - 0.4425$$

If μ (Comprehension of each unit of information) is even as high as 0.8 (80%) then pattern strength (Δ) is only 0.3575 and the true picture of just how far from reality our comprehension might be slipping becomes apparent.

However, salvation is possible, in one of three very special ways, assuming that these individual values cannot be improved upon.

The first salvation is the consequence of degree. Remember that the pulsonic model of our quantum eventuality universe has revealed an infinite number of finite elemental factors. The universe is a place of degrees and shades of being, not hard-edged opposites. Many accepted notions of an adjectival nature are simply couched in poor phraseology, and thus lead to fuzzy conceptions.

Hot and cold, for example, suggest that there are two thermal properties that are straightforward opposites of one another. But the reality is that a single property exists which is perceptible in degrees, artificially divided by our perception according to its level or effect. The single property is thermal agitation, and what we subjectively attach the label 'cold' to is in fact merely a lower level of thermal agitation than the one to which we subjectively attach the label 'hot'.

Heaviness and lightness are another example of a single property – mass - divided artificially into opposites according to how we determine the correct mass for a particular object. A car that weighs half a tonne in Earth gravity would be considered light, but a man who weighed 20 stone in Earth gravity would be considered heavy, even though he is notably lighter than the car. A feather is not 'light'; it is simply an object containing little mass.

Long and short, Tall and short, clever and silly or creative and destructive are similar instances of single properties divided artificially by our relentless pursuit of differentiation. But the fact is that a short person is not an opposite of a tall one, merely an example of someone with a lesser quantity of the property 'tallness'. You might wonder how the 'Consequence of Degree' can help us achieve better results for 'view-eventuality' and reduce the losses at our 'primary phase in entropy's degradation of information'. It is this; when a subject is studied intensely for a considerable amount of time by a hard-working mind the Observer's Stimulus Potential can exceed unit value and thus compensate for entropy. This is usually the case where genius is concerned, and it comes about because the degree of factor A^0 exceeds unit value. If we return to our first set of figures but replace the value of A^0, we obtain a significant improvement;

$Q^1 = 0.9$

$q_0 = 0.215$

$A^0 = 1.35$

$O^{sp} = 1.161895004$

$$Q_i^a = \frac{Q^1 + q_0}{2} \text{ x } A^0 = 0.752625$$

$$_vQ = Q_i^a \text{ x } O^{sp} \text{ x } Q = 0.874471227 \text{ x } Q$$

The second salvation comes through the relationship of quantum eventuality to MID and MAP levels, which must naturally have potential reflections in the QIA values of 'view-eventuality'. By paraphrasing the MAP calculation we can arrive at an equivalent maximisation of a Quinarc valuation; $_vQ\,c^3$. Unfortunately the flip side of this is the comparable presence of an analogue for MID; $\sqrt[3]{\dfrac{_vQ}{c}}$. If $_vQ$ was 2578 then $_vQ\,c^3 =$

$6.946166743 \text{ x } 10^{34}$ whilst $\sqrt[3]{\dfrac{_vQ}{c}} = 4.413881977 \text{ x } 10^{-3}$, which are clearly not within the realms of normal consciousness. One value is so high it exponentially exceeds the available information of its source, the other so low that we are looking at a consciousness as feeble as a pebble. Fortunately the mind does not operate across such a wide range of values, which are thus clearly off the scale for QIA calculations. This is mainly because consciousness is a low energy, high efficiency activity of matter and the relativistic function c^2 plays no part in it. This should be no surprise, since c^2 is already bound up with QEP and thus cannot be counted in further extrapolations. Even with the QEP analogue itself - $_vQ\,c^2$ - the result is too high to lie within the capacity of consciousness. However, the answer to QIA derivatives of MAP and MID is mercifully straightforward; our MAP equivalent is MIP – Maximum Interception Potential.

$$\frac{_vQ}{1 \div \sqrt[3]{_vQ}} = {_vQ} \div \sqrt[3]{\frac{1}{_vQ}}\text{ , yielding 35349.00693 for the example above.}$$

Our MID equivalent is MRI – Minimum Residual Interception - $\sqrt[3]{_vQ}$ which gives us a figure of 13.71179477 for the above example; far more understandable.

I propose that any conscious mind – human, canine, equine or non-terrestrial – has the MIP and MRI values as its upper and lower universal participation levels. They represent the possible range of conscious activity with respect to a particular stimulus, from a level so nominal that it appears to be hardly aware at all, to a level that exceeds and maximises available information.

The Identity of Knowledge

Axiom 1:

Knowledge is the awareness of factual or fictional elements of information. Information only conveys potential knowledge, in that it transmits a hypothetical description of reality, which may or may not be witnessed by consciousness and which, even if witnessed, may or may not be interpreted accurately. Knowledge may thus be a pure or impure awareness of information and can only exist when transmitted information is received, accepted and understood by consciousness. Knowledge cannot exist when;
1. Information is not transmitted
2. Information is transmitted but not received
3. Information is transmitted and received but not accepted
4. Information is transmitted, received and accepted but not understood.

Because knowledge can persist in the memory it does not require continual transmission from the source, thus increasing the efficiency of consciousness. Indeed, it is a property of consciousness that knowledge can be memorised and applied in a variety of situations thereafter, and a property of intelligent consciousness that knowledge is actively sought for the improvement of all situations encountered.

Axiom 2:

Knowledge can only occur as a result of transmission, or as we prefer to call it, communication. In order for knowledge to evolve in a unit of consciousness there must be an efficient medium through which information is transmitted. When the transmitter is a unit of consciousness itself the medium involved is most commonly a language of phonetic, sonic, visual or other sensory symbols. When the transmitter is a unit of unconsciousness the medium involved is direct expulsion of information-carrying micro-photons that impact with the receiver's senses and may amplify by logical extension the receiver's existing knowledge. The achievement of knowledge therefore depends on successful combination of transmission clarity, communicative fluency, an efficient host medium, essential stimulus sensitivity and interpretative ability in the receiver.

Axiom 3:

Rationality seeks precision in knowledge. Information broadcast by units of unconsciousness is expelled in a continuous but random stream, so that when it bombards consciousness it does so in a general and unselective manner. On the positive side this information is unambiguous because it is generated by absolute reality, but knowledge is ambiguous because it derives from reactions to received information. Systematic, selective and comprehensive reasoning help a unit of consciousness to improve the validity and accuracy of these reactions, in order to close the gap between unambiguous fact and ambiguous thought.

Axiom 4:

The equations above help us to track the discrepancy between fact and thought and are the primary interaction calculations documenting the interplay between the universe and awareness. Entropy, Quantum Weirdness, Quantum Orthodoxy, Observer Stimulus Potential, Observer Activation, view-eventuality, Maximum Interception Potential, Minimum Residual Interception and reality itself all interact even at this primary interface.

In documenting the identity of knowledge we need to run a brief analysis of consciously selected subjects showing how different spheres of knowledge are employed to answer questions concerning reality.

Subject Phase	Prime Members	Aim-answers
OBSERVATION	Geography	Where are we?
	Earth Sciences	What surrounds us?
	Physics	What systems define physical reality?
	Chemistry	What systems define chemical reality?
	Atomic sciences	What is the heart of existence?
	Astronomy	How much are we a part of?
	Mathematics	What relationships exist?
	Biology	What are we?
EXPERIENCE	History	When are we?
	Archaeology	What/who preceded us?
	Anthropology	Who else is there?
	Sociology	How are they?
	Psychology	How am I?
	Art	How can we depict reality?
	Literature & fiction	How can language extend reality?
	Music	How can sound reflect reality?
WISDOM	Ethics & morals	Is our behaviour considerate and true?
	Logic	Is our thought reasoned and justified?
	Philosophy	What is our purpose?

Answering these questions, or even merely attempting to, bring us closer to reality. Our identity is raised against that of the subjects we study and master, thus reducing our inequality with that information. By pursuing the acquisition of knowledge and the transformation of information into knowledge we increase our link with reality and increase our participation in the structure of the macrocosm.

Our grand universal structure, modified from Section A then, is as follows;

Macrocosmic information

|

X-cosmic information (unavailable)

|

Photo-cosmic information (available)

|

The Subjects of Knowledge

|

Units of Consciousness Knowledge

The Quantum Interception Arc of Civilisation

The concepts of civilisation – patience, compassion, responsibility, reverence, tolerance, consideration for other life, pacifism, vegetarianism, mercy, politeness, self-education, self-discipline, altruism, conservation, virtue, empathy, logic, harmonic co-existence with nature, integrity, honour and reason – are overwhelmingly dominant in a civilised species. Thus the identity of a civilised species is raised level with that of the elemental factors of civilisation noted above. And the Quantum Interception Arcs of these civilised concepts are pronounced in favour of higher values (nearer to unity) within every unit of consciousness within that species.

In a civilised species any uncivilised element would be extremely rare; it is only the civilised conduct of a being or species that can confer upon it the status specified by the unpleasant term 'superiority'. It is precisely the civilised being and species that will recognise its own failings and seek to eliminate them, never resting until they are reduced to a barely noticeable level. As such the placing of itself upon a pedestal – supposedly above other beings and species – is an action that is incongruous with civilisation, for it proceeds from the antithesis of the elemental factors of civilisation.

Acts of barbarism and anti-congeniality would be out of place and conspicuous in civilisation, just as civil conduct would be rare occurrences in barbaric societies or individuals. Just as the Quantum Interception Arcs of knowledge factors are proportional to conscious affinity, so the hallmarks of civilisation are proportional to conscious affinity. Paradoxically, it is the being or species lacking the civilised refinements listed that will place itself upon its own pedestal, unaware of its own failings and thus promoting them, and thus looking down upon anything that is not it. By doing this it will unwittingly confer upon itself the status specified by the unpleasant term 'inferiority'.

Where can we place ourselves within these definitions? It is clear that there are acts committed by humans, both individually and collectively, that disqualify us from any definition of civilisation. We are not guilty of rare breaches by a minority of our population; instead we breach the minimum requirements of civilisation regularly and in abundance. With each additional breach we further disqualify ourselves, adding insult to injury and tarring those who try to be civilised with the same brush. We have a long way to go before we achieve a respectable Quantum Interception Arc for civilisation.

Entropy and Knowledge

If information and entropy work against each other it follows that knowledge and entropy work against each other, because knowledge is the conscious activation of information and is structured. Over time entropy degrades knowledge; the results are forgetfulness or amnesia. A conscious mind that seeks to reject information, avoid knowledge and promote entropy acts only from ignorance and chaos. I have said before that entropy is the enemy of life and intelligence is the enemy of entropy; thus intelligence is the ally of life and destructive behaviour is the enemy of intelligence.

Section F: Multiple Wave Expansion

Principles of Expansion Eventuality

The basic measurement unit of this theory is the Quantum Expansion Arc or Quexar, which measures the loss of clarity when reality expands to consciousness. It is a far more subtle measurement than the Quantum Interception Arc, and is the realisation of six uncertainty principles;

1. (Heisenberg's) The velocity and position of a particle cannot be measured simultaneously.
2. T, V and M cannot be measured simultaneously, in any combination.
3. One cannot accurately measure T, V or M at the same time as movement.
4. Emission and absorption cannot be measured simultaneously.
5. New isofluxes within a finite event cannot be created simultaneously, whilst new isochrons across the macrocosm cannot be created simultaneously.
6. A temporal edifice cannot be both measured and described simultaneously; description must follow measurement and whilst description proceeds measurement may alter.

The central core of this theory is that reality as a transmitter of information obeys a quantum expansion arc that continues to expand until information degrades to zero. The process of transmission follows this arc in an insentient manner, oblivious even to the presence of Quantum Interception Arcs intersecting it; intersections of QIA do not deplete or affect the QEA at all. A Quantum Expansion Arc represents a loss of clarity due to entropy, amplified in ratio to the perceptiveness and receptivity of a receiving unit of consciousness. The further away from reality a unit of consciousness is the greater its distortion of reality will be. Since entropy works through time it follows that as time progresses the further from reality will knowledge become. Even if no QIAs intersect a QEA the ultimate destination of the QEA is zero-base and so the depletion of information and knowledge.

Distance from reality can be time driven, by entropy introducing chaos into an organised system. Depletion of information is one way in which to measure the passing of time. I also mentioned a unit of consciousness being far from reality; the greater its distance the greater its internal distortions of reality. The distance being considered here is not strictly physical, although it is true that knowing what is happening somewhere becomes increasingly difficult with distance. I am living in England at the moment; on a daily basis I see only a part of it. The probability is that I have access only to that part that comes under my scrutiny, and only whilst I am present in a particular vicinity. This blind spot of uncertainty multiplies as the radius of our concerns expands, and it is precisely this dilution that QEA focuses on. The real distance, however, is in attitude.

Knowledge depends not merely on physical proximity to the bits of information that create facts, but even more importantly on cerebral proximity. It is ignorance, and more horrifically the desire for ignorance, in a so-called intelligent species that distances consciousness most conspicuously from reality. No single mind can contain all the facts of reality, so it is a deeper and more arcane problem than mere lack of information; any mind can be forgiven for not having acquired sufficient information to build a working knowledge of certain aspects of reality. Even the most gifted, the most comprehensively schooled intellect will have plenty of blind spots. What is difficult to forgive is that within a species that claims to be civilised there are collections of consciousness that embrace ignorance, that turn away from knowledge, that seek intellectual vacuum and court irreverence for existence. Since it is only knowledge (and hence information) that connect consciousness to reality, such rejection is self-destructive.

Einstein knew both the limitations and importance of knowledge. It should be realised that whilst someone in command of a body of knowledge – say, chemistry – may miss discoveries and understandings within that subject, such possibilities of lapse increase exponentially with the absence of chemical knowledge and experience. It is similar to the situation of driving a car; an experienced and gifted driver, even one who has trained specially to cope with extreme driving conditions, can have an accident, lose control in a split second of misjudgement and end up in serious trouble. But removing driving experience and knowledge will only increase the element of risk. True, an inexperienced driver with only basic skills may repeatedly drive safely, but that is more from luck than skill.

It is mainly the seriousness we apply to knowledge that determines the extent of our participation therein, and more importantly the value. The advanced driver seeks pro-actively and with zeal the skill of controlling a motorised vehicle, the chemist seeks in a similar manner the skills of chemistry. Certainly one might find that even with application a subject eludes comprehension; affinity for it is slight. But application is important even if ultimately pursuit of that subject is abandoned to save face. There is no dishonour in opening a book about counterpoint, only to be forced after the first twenty pages to close it firmly and return it to the book shelf in humbleness and awe at the gift of Bach. There is dishonour in possessing contempt for the information contained in such a book, or Bach's superb gift.

It is in trying and seeking to master knowledge that consciousness shows its true power and connection with reality. Intelligent participation is more the desire to participate and learn than a scorecard of the personal triumph of memory over fact. What is important is not that we failed to learn something specific, but that we failed to want to learn at all. Our collective history is not particularly encouraging with regard to our ability as a species to improve after witnessing and analysing our previous mistakes, then determining how to correct the failings within ourselves that lead to those mistakes. Thousands of years ago great Greek philosophers commanded levels of logic, pure reasoning and perception that make today's strongest thinkers appear weak in the head. What has gone wrong?

It is our attitude that has gone wrong. More than anything else attitude determines whereabouts on the Quantum Expansion Arc our Quantum Interception Arc creates its junction; towards the stronger end of the expansion or nearer to the entropy depleted end? The philosophers of ancient Greece loved thinking about life; it is no accident that 'philosophy' begins with 'philo', meaning love. Their attitude was positive, intelligent, alive and refined. Today there is a reaction against such noble proclivity. Anyone displaying it is in danger of being labelled with derogatory adjectives; the prevalent attitude of the present favours entropy, a wasteful descent into disorder. Chaotic sounds from those with little or no musical ability fill the air and chaotic thoughts from those with little or no honourable thought fill books that have no constructive mind-building purpose. In every field the majority is carefully primed to crave chaos, by unscrupulous companies and individuals that can only succeed by taking advantage of collective obsessions in order to make huge sums of entirely undeserved money. Chaos is easier to sell than order; it requires little or no intellectual effort. But to favour chaos over order is shamefully without merit.

In its mentality the human race is now causing the entropic decline of knowledge to accelerate; we are forcing the QIA to splice the QEA further along in its demise. Our race for the future and away from the past is frenzied and disorganised because it is poorly judged. Instead of clearing up the savagery and failures of history and memory we continue to carry them with us; war, violence, crime, abuse of other species and dishonesty - activities that ensure our continued stupidity. But that of the past that was strong, constructive, character and mind building – the music of the great composers, good manners, honour, reverence and discipline - we seek to abandon as 'not trendy'. Such behaviour is dysfunctional.

In England we have been arguing about anti-fox hunting laws; in a civilised society such arguments would not occur because there would not be anyone low enough or sick enough to accept the commission of any atrocities, let alone desire them. Laws are required only where people cannot be trusted to behave decently, and as our society 'progresses' more and more laws are needed, suggesting that our behaviour is worsening, a clear illustration of social entropy.

The expansion from reality to consciousness demonstrates how clarity is diminished as it is stretched across time and space. The Quexar measures the point at which conscious interception occurs, and therefore the remaining integrity of the expanding information. It is what is left of reality when we happen to notice it.

Our classifications for determining Quexar values are;

$$Q_E^A = Q_i^a - \emptyset T = \text{QEA in Quexars}$$

$$S = \text{success rate of QEA} = \frac{Q_E^A}{Q} \times 100$$

Remember that the universal Quantum Weirdness equation is;
QW = Q x rf
Where rf = reduction factor as % deviation from Quantum Orthodoxy / 100
And the general equation for Quantum Orthodoxy is;
QO = Q – QW = Q x (1-rf)

For the purposes of correctly calculating Q_E^A we need to build quantum weirdness into the equations for quantum expansion. Thus where β = 1-rf = broadcast efficiency and R = reception efficiency, first substituting Q_E^A for Q in each calculation;

Potential of transmitter = βQ_E^A and Potential of receiver = RQ_E^A

$_pQ$ = para-eventuality = $_pT\,_pV\,_pM$ (in proportion of TVM) = βQ_E^A x RQ_E^A
Values of Q are taken as decimal fractions of unity (1).

Applying these equations to an example based on Q = 1, T = 0.5 (50%) elapsed seconds, Entropy = 0.5 then; $\emptyset T = 0.5 Q_i^a$ and $Q_E^A = Q_i^a - \emptyset T = 0.5$

If $\beta = 0.8$ and R = 0.7 then $\beta Q_E^A = 0.4$, $R Q_E^A = 0.35$ and $_p Q = 0.14$

We can continue the merging process by combining QEA with QIA;

$Q^1 = 0.9$
$q_0 = 0.215$
$A^0 = 0.88$
$O^{sp} = 0.938083152$

Then $Q_i^a = \dfrac{Q^1 + q_0}{2}$ x A^0 so $Q_i^a = \dfrac{0.9 + 0.215}{2}$ x 0.88 = 0.4906

$_v Q = Q_i^a$ x O^{sp} x Q = 0.4906 x 0.938083152 x Q = 0.460223594 x Q

Q is no longer a value of 1; it is replaced with $Q_i^a = 0.4906$ and substituted into the above calculations.
$Q_E^A = Q_i^a - \emptyset T$ and $\emptyset T$ being 50% of Q_i^a (T = 0.5) = 0.2453

If $\beta = 0.8$ and R = 0.7 then $\beta Q_E^A = 0.19624$, $R Q_E^A = 0.17171$ and

$_p Q = 0.03369637$

It may seem that each level of information transmission into consciousness produces ever more worrying results, but remember that trying to download information from infinity into limited finite units is never going to produce high yields, therefore 0.03369637 is more respectable than it may at first appear. It does illustrate, though, that our ability to comprehend is restricted and that there is an uncomfortable variance factor that should rightly make us keep an open mind about our perceptions of the universe around us.

Para-eventuality ($_p Q = {_p}T \, {_p}V \, {_p}M$) is our 'second phase in entropy's conscious degradation of information'. Now, more clearly than ever before, the reduced strength of reality refers to what is potentially storable in a unit of consciousness. It is an approximation that proceeds to a greater or lesser degree parallel to reality. The more facts a mind absorbs and the more strenuously it thinks, the closer its inner world is related to the outer world. We must keep in mind that there can never be total equivalence, what consciousness realises may not necessarily be correct and no matter how powerful or accurate a mind is the capacity for quantum weirdness and degradation is a chasm that cannot be ignored. It is the quantum dispersal of information across the chasm of the Expansion Arc that gives this theory its prime component. It is 'Multiple Wave' because of the inevitable intersection with the Interception Arc.

All is not lost, however. As we saw in the analysis of Quantum Interception Arcs, there are ways of compensating for the deficiencies of sentience. Thought is the most efficient use of energy, being exponentially less energy intensive than physical effort yet exponentially more result intensive. And intelligent thought is even more streamlined than that. A well run, well balanced, positive, organised, efficient, potent and refined sentience can overcome the negating factors of information transmission and management, and arrive much nearer to reality than would otherwise be possible.

The key value is in the function R. In Quantum Interception reckoning it was the values of A^0 and O^{sp} - respectively talent (activation of observer in relation to a temporal edifice) and reverence (observer's stimulus potential) - that improved the results. In Quantum Expansion R is simply a sentient being's IQ.

Prediction and Predetermination

If there is an explosion, what are the chances that a particular person will be hurt? The proposition contains many random elements, of which the most influential is the actual chance of an explosion occurring. This is the combination of several random factors.

A. The presence of sufficient volatile material to cause rapid catastrophic disassociation of itself and proximate stable material.
B. Lack of sufficient containment to prevent such rapid catastrophic disassociation.
C. Sentient incompetence in failing to provide such containment or prevent the accumulation of a critical quantity of the volatile material.
D. Sentient iniquity in deliberate accumulation of a critical quantity of volatile material, and simultaneous deliberate removal of adequate containment.
E. The presence of a catalyst sufficient to trigger disassociation.
F. Sentient incompetence in permitting such a catalyst to come into proximity with the volatile material.
G. Sentient evil in deliberate combination of catalyst and volatile material.

Assuming that the conditions in A to G above are precisely arranged so that an explosion will occur either through negligence, incompetence or iniquity, the remainder of the proposition continues to rely on random or seemingly random elements. The probability that a particular person will be hurt in this explosion depends on more chance circumstances;

1. The timing of the explosion coincides with a time frame in which that person is present inside the perimeter of the blast area or danger zone.
2. There is nothing within that perimeter, of sufficient strength and in a suitable position, to act as a protective shield for the unfortunate victim.
3. The extent of the damage sustained by the victim will depend on particular distributions of force and weaknesses within the volatile material, proximate stable matter and the victim themselves.
4. The location of the explosive matter will depend on chance elements like decisions made by incompetent or iniquitous sentience.

There is no such thing as the 'bullet with your name on it'. Random incidents lock together bit by bit until there is a point of no return. That point of no return is the consequence of a precise series of random incidents that act cumulatively to prevent change.

Consequence is a different matter from predetermination. Something that happens as a consequence of a previous series of circumstances that were once random cannot be called definitive predetermination, because it is only the exact grouping (and in particular the final addition) in the chain that causes the lock. At any point up to that final element the chain could have been undone or set quite differently; concrete needs time to set and when it does it is difficult to alter.

Our explosion, therefore, was brought about by the interactions of A to G and 1 to 4. A change at any point could have caused a different outcome. Predetermination and determinism in general are clearly inadequate concepts, or at least they are incapable of properly defining reality. Events of the future are not rigidly fixed in advance and remain partially unfixed until the present has unfolded to touch the future. This means that the idea of complete predictability is also without merit or foundation. The validity of prediction only approaches any strength when the gap between present and future has closed to within arm's length. Logic alone tells us that determinism is not feasible; if the universe is predetermined then the future is precisely predictable, and therefore precise action can be taken to alter its course. A clear paradox is indicated by such an assumption.

The opposite of determinism is also a paradox; in a universe where every unfolding eventuality is so unruly that no degree of expectation is stable, action cannot be taken to alter specific outcomes. Therefore the future would be simultaneously unpredictable and unalterable, a clear contradiction in terms.

Reality must be a compromise between rigid fixture and chaotic maelstrom, but exactly what manner of compromise? Do we have a limited degree of choice in a strictly predetermined universe? The London Symphony Orchestra had been booked to travel on the maiden voyage of the Titanic but fortuitously their concert plans were changed at the last moment. They had to travel later on the SS Baltic and thus all that musical talent was saved. Clearly the universe cannot be too strictly predetermined, nor choices that limited.

The next compromise that comes to mind is that we have considerable choice in a loosely predetermined world. It sounds more likely but is it the case? It cannot be, because it does not take illness into account, or whom we have as family. There are two further options that can be discounted immediately and without the need for investigation, primarily because their absurdity is immediately apparent; we have no choice in an undetermined universe or we have complete choice in a strictly predetermined universe.

Our only way out of this little conundrum is to return to our realisation of consequence and apply it to Quantum Eventuality from the ground up, so to speak. According to the uncertainty principles at the beginning of Section F, quantum fluctuations in reality have a fuzzy randomness that creates an unpredictable nature at the quantum level. Events in the SNPWC, which largely revolve around the exchange and interaction of micro-photons, do not react easily with major events and are thus difficult to manipulate or influence at non-quantum levels. In the realm of the SNPWC events are carried by the micro-photon, which exclusively controls relativistic transformations.

Outside the SNPWC, where micro-photons are transmitted through vacuum (and are thus temporarily detached from the reaction medium of both dynamic and passive fundamentals), relativistic transformations become increasingly more difficult the more dilute the CSFC, and hence its Cosmic Event Tension, becomes. The SNPWC thus only submits to local micro-photon transfer and changes are governed by relatively minor quantum fluctuations whilst being left undisturbed by major cosmic events.

At the same time, major cosmic events do not interact directly with the SNPWC; they require other incoming major events to become influenced. These abstract and generalised definitions cover all primary contingencies and therefore lead us to the truth of Consequential Determinism, Contingent Expansion and Provisional Locking. We occupy a universe that is fairly predetermined on a large scale but virtually undetermined on a small scale. The smaller the scale becomes the less determination is evident. This means that choice becomes more limited as scale increases.

Consequential Determinism is in principle very like inertia; the greater the mass is the more difficult it is to alter its state and direction of motion, including 'at rest'. But with Consequential Determinism the inertial factor is time; the greater the mass the more its motion through time locks in place. Relativity Physicists talk of the cone of time, and it is the direction and length of that cone that Consequential Determinism defines. As the cone expands its destination and terminus become ever clearer and more focused.

The sub-mechanism is Contingent Expansion; it has been noted that consciousness prefers interception before entropy can degrade information. The Quantum Expansion Arc measures interaction between reality and thought and Contingent expansion is directly related to the QEA. The intersection point of QEA and QIA illustrates the Contingent Expansion of Consequential Determinism at which perception intercedes, and the degree to which consciousness can bring about change in the temporal edifice being observed.

Provisional locking indicates a projection of the temporal course of an event object, based on its current history and expansion arc. As mentioned earlier, the further along in its own history an event object is the more securely it locks into a particular outcome, until a point of no return is reached; after this no matter what is done to the event object its destiny is irreversible. This very beautifully and very symmetrically reflects the event object's condition in entropy. At inception it holds greatest temporal coherence; the fullness of life stretches before it and prediction of specific events to come is almost completely impossible. As it proceeds along its Quantum Expansion Arc it loses information as entropy approaches maximum. When the QEA is intersected by QIA only the object's degraded evolution is discernible to consciousness. This means both the future and past must be extrapolated by conscious endeavour, one from reconstruction and the other from pre-construction.

All is still not lost, though. The chasm between reality and conscious perception of it may be widening with each new revelation of the relationship but we have already seen how knowledge, intelligence and efficient perception can help bridge the gap to a certain extent. It is a sobering thought to remember, however, that no matter how strong the bridge we build across that chasm, it is always possible to fall off with the folly of a misplaced foot, or for the bridge to give way.

Determinism and the 'not I'

If x = "the sum of the 'not I'", then x exists with or without the participation of the 'I'. In this sense the observer's inability to prove or disprove the existence, nature and substance of x is a quality of para-determinism. It does not mean that the 'not I' is unreal or fundamentally different to image and observation, or that reality is observer created. Indeed, if reality were observer created we would be living in a universe that was simultaneously super-determined and super-undetermined; our observation of reality would alter it to accord with our perception making it both precisely controllable yet incoherent.

The truth is that x = "the sum of the 'not I'" exists as it pleases and our intervention, participation and existence within it make little difference, reinforcing our previous notion of a universe that is fairly predetermined on a large scale but virtually undetermined on a small scale. The main consequence of this is that we may not be able to fundamentally alter reality merely by interacting with it. We must engage in specific interactions that possess specific reactivity with the situation in question, and we may not be able to prove the results afterwards because part of reality will always be unknowable.

If someone throws a pencil from a window they may predict where the pencil will come to rest, using the force, trajectory-angle, wind factor and mass in their calculations. But they cannot prove their prediction is valid, or that their observation of the actual resting-place is accurate. Even if the pencil ends up precisely where they said it would, there is no real way of categorically attributing this to clairvoyance or scientific precision. There are too many factors along the Quantum Expansion Arc that could contribute to the outcome and only when the pencil loses all momentum can any reasonable description of its whereabouts be given. But by this time the pencil is far from the pencil thrower; intervening distance can fool the eyes, possibly even into not seeing the pencil at all.

Because of the separation between the 'I' and the 'not I', there is a vast gulf between the true order of cosmic reality and an observer's perception. This gulf is more than sufficient to thoroughly destroy any possibility of observer-created reality. We can only meaningfully say that we believe or assume, never that we know, unless we are in possession of a huge body of fact, a proven understanding of the patterns behind that fact and an unquestionable motivation of logic. For, if it were possible to know and understand the exact state of every subatomic particle in a particular temporal edifice, we could theoretically comprehend every possible energy exchange between those particles. But does that mean we would actually be able to ascertain which of those energy exchanges were going to occur?

Armed with such impressive awareness could a person precisely predict the edifice's possible condition in say fifty or a hundred years' time? Note that I say 'possible' condition; even with such an amazing capacity a seer or God could not know what outside agencies may enter the reaction arena of the temporal edifice under consideration, thus bringing about changes beyond the possibilities discernible in isolation. And nothing in the universe can remain isolated indefinitely, not even something undergoing relativistic transformation. Plus, we must bear in mind that once a temporal edifice undergoes relativistic transformation it is effectively disconnected from the rest of reality and thus no longer observable.

So, even an expert cannot know all things about the subject of their expertise. This is as true of the subjects of knowledge as it is of concrete objects. An expert in the field of music, for example, will know every subatomic particle (piece of information) of the subject, but will still not be able to predict what future musical compositions will be like, even of their own writing. It is this inevitable lack of information (both through the secretive essence of reality and our inefficiency in comprehending it) that causes the indeterminate nature of the universe. Clearly prediction is insecure, possibly fraudulent, and gullible acceptance of it is folly.

Outside of us the cosmos is reasonably self-determining; it knows what it is doing and we must be clear that our inability to see and measure exactly what reality's plans are is in no way the fault of an indeterminate universe. It is our failure, not God's. The main trouble is that consciousness works in the void between the transmitter and receiver of pieces of information. It is by no means certain that any transmitter can be expected to be 100% efficient in its transmission, because of entropy. And in our case the receiver is a combination of our senses and our consciousness. Entropy and other limitations ensure that we can certainly not trust ourselves to be 100% efficient receivers of information, and most definitely are not even consistent at any level. We are quite unreliable. The main pitfalls are;

Transmission

1. Weak output through entropy, system failure or system inadequacy
2. No output at all
3. No output at frequencies compatible with the receiver
4. Output not energetic enough for recognition
5. Output too hectic, rapid or energetic for absorption
6. Output from either inept or deceitful consciousness

Reception

1. Sensory failure through design incompetence or entropy
2. No reception at all due to absence of sensory equipment
3. No reception due to broadcast frequency being incompatible with reception
4. Input not energetic enough to register
5. Input too energetic, fast or chaotic for controlled absorption
6. Input from an incompetent or deceitful source
7. Consciousness incapable of efficiently deciphering sensory input
8. Consciousness deceitful with itself
9. Consciousness not linked to sensory equipment
10. Consciousness incapable of logical analysis

If anyone doubts the above they only have to consider that after 10,000 years of supposed civilisation the human race is still actively involved in destructive behaviour towards other life. There are, for example, the matters of de-forestation, pollution, the murder of sentient beings (even individuals from amongst our own kind), the commission of other (non-fatal) crimes and the commission of other (non-fatal) cruelty, abuse and wastage. All of these failings usually fall under section 10 in the list of reception problems, but sometimes proceed from other problems in the list. The universe teaches us many things but we seem to have trouble learning, both individually and as a species. When humans talk about how advanced and intelligent humanity is it is sometimes difficult to avoid either crying or laughing.

Probability and Possibility

Most of this theory has been spent considering the fine points of reality, from a level beneath quantum particles to one beyond the photo-cosmos, from billions of aeons before the Big Bang to the future and beyond. Moving through time, space, matter, relativistic connections and thought we have constructed comprehensive models and extracted detailed elements of eventuality. Presently we have been occupied with the interplay of these elements and models relevant to prediction and predetermination.

The accumulated evidence points quite clearly to an inviolable forward flow of time and a clear relationship of cause and effect. Prediction becomes slightly more robust as mass increases and entropy is exhausted; at the inception of a temporal edifice prediction concerning its future is virtually impossible because no information has been leaked out by the Resonance Excision of micro-photons. Near termination the greater bulk of information that can be lost has been, thus it is easier to determine what will happen because the future possibilities are more firmly set in their course. Temporal inertia has set in. However, it is more difficult to look back along the cone of time and determine precisely what previous events lead to the edifice's current configuration.

More than anything else, cause and effect is a sequence of steps from beginning to end – our chain of Consequential Determinism – that begins with high profile history/low profile future and moves inexorably to low profile history/high profile future. In SNPWC terms the Contour flattens with time, as its reserve of excised micro-photons dwindles and becomes less energetic. Consciousness generally prefers the time of inception because greater information is readily available within the edifice itself, but the opportunity cost is the inability to make accurate predictions. At the end of a temporal edifice's chain of Consequential Determinism, prediction becomes easier but the opportunity cost is that most of the events that produced the chain have passed into history and much information has been lost. Contingent Expansion where perception intercedes near inception of the edifice permits us to exercise greater control over the future but in a blind way; with a Contingent Expansion where perception intercedes near termination we have less control over the future but plenty of sight.

This may explain why 'new' things are generally preferred to 'old', but it does not explain the motivation behind the desire of consciousness to intercede in the progress of a particular edifice at a particular point in its chain of Consequential Determinism. For that we must consider the realm of Provisional Locking; probability and possibility. And we should not ignore the considerable presence of logic either, for cosmic reality is above all other characteristics logical. On the other hand consciousness is distanced from cosmic reality and struggles to achieve logic, not always very successfully.

As already observed, Consequential Determinism arises from Contingent Expansion and Provisional Locking. Only the past and present are set, their isochrons and isofluxes spent forces. In fact we can now provide accurate mathematical definitions of past, present and future. The past is all that has happened before the extant isochron, the present is at the cusp of it and the future is all that stretches beyond in the realm of the unknown. When the isochron crystallises it recedes into the past and is replaced by a new one. In an infinite universe there is no end to those that have gone and those that are to come. We can only partially extrapolate the contents of the future, based on historic trends and ongoing Provisional Locking. The accuracy of our extrapolation depends on logic, probability and possibility.

Previously I mentioned the fallacy of opposites; it is a condition of human mentality that adjectives like 'hot and cold' are seen as descriptions of separate and opposite qualities. Scientifically speaking there is only one phenomenon – thermal agitation – so that the adjectives 'hot and cold' in reality describe subjective interception points along a single arc. The same is true for logic, probability and possibility. Each property has its own Consequence of Degree.

Are there degrees of impossibility? This question is addressed first so that we can eliminate redundant functions from a working model. Impossibility is that which cannot exist because it is excluded from the universe. It includes impossible substances, impossible structures for existing substances and impossible events for existing structures. Since there cannot be degrees of exclusion, non-existence or any quality supposedly maintained by something that does not exist, there can be no degrees of impossibility; all impossibilities are equal.

Even the wildest excess of imagination is not excluded from existence; whilst imagination occupies a tangible corner of existence it is only the fanciful products of imagination that may lie beyond possibility. Remember that we are concerned with aspects of existence and reality, of which imagination is a member by virtue of its growth from consciousness. The bizarre creatures that inhabit imagination are the non-realities and impossibilities, possessing no Consequence of Degree and no hierarchy. Impossibility will always remain fictional, like the juxtapositions in surreal paintings, but sometimes invention can become reality.

Could there be such a thing as a star made of ice cream? No, because the forces and quantities in even a small star preclude the presence of complex chemicals, especially of organic variety, required for ice cream's existence. Furthermore, we are talking about a non-natural substance of human contrivance, being amassed in a quantity equivalent to several hundred planets at least. Clearly this is not a construction project possible from the resources of even 1000 planets.

Is it possible for me whilst I am in England to hold the hand of a lady whilst she is in the Ukraine, or kiss her for that matter? No. These two examples illustrate impossibilities of structure and event, but what of the impossibility of substance? This is more of a problem, since the only atomic frameworks we know are the ones in our local photo-cosmos. We must be more general and theoretical in such an example; as far as we know it is not possible for a substance to be both gaseous and solid simultaneously, or to be both liquid and self-shaping simultaneously. There may be other examples, but you get the picture, don't you?

We have discounted impossibility from both existence and the Consequence of Degree, which leaves all of reality as possibility, whether potential or actual. Probability, then, is the likelihood of discovering a particular event or edifice either in the realm of the potential or the actual, and is therefore very much a Consequence of Degree. Our picture so far is a little clearer;

1. Degrees of possibility
2. Degrees of probability
3. Degrees of logic

How do causality and chance interact with these three Consequences of Degree?
Let us suppose that a person who earns £x per year wishes to buy a particular new car that costs £2x. If this hypothetical car buyer has ≥£2x in savings then the possibility that they can buy the car outright is extremely high; the only potentially reductive factor is their will. If they have ≤£2x in savings then the possibility of buying the car outright is reduced, but the possibility of buying in some combination of cash and borrowing is still high. The possibility can be further reduced, however, if the person's salary is low, they lose their employment or their savings are needed for some other purchase. Other factors that might alter the Consequence of Degree are either good or bad publicity concerning the car in question, availability of the car or the launch of another make or model that captures the buyer's attention even more.

This specific scenario only becomes impossible in the following circumstances;
1. The buyer's attitude toward this car or its manufacturers becomes so damaged that they would not purchase the car even if it was heavily discounted.
2. The buyer's financial situation substantially deteriorates, making the car no longer affordable.
3. The buyer becomes influenced by environmental arguments that persuade them to cycle, walk or keep the car they have already.
4. The buyer dies.
5. The buyer becomes unable to make decisions through mental decline.
6. The car is withdrawn from the market.
7. The buyer is barred from driving and owning a car.

In determining degrees of possibility we are really looking at the force of chance or random elements; the strongest connections of random elements are the ones creating the more 'possible' chances. Where random elements are incapable of connecting with one another, the realm of impossibility has been entered. In a way this is our phase of Consequential Determinism. The strength or presence of a random element, or collection of random elements, creates the temporal inertia necessary to carry a specific event forward. The stronger the connections of the random elements in an edifice are the more they are locked into time's arrow and the more substantial must any disruptive force be to 'unlock' them. At the start of the Quantum Expansion Arc possibilities remain at their greatest range and diversity. What time does is to strengthen the connections by narrowing the cone of possibility as the cone of time opens. It becomes easy to see that it is the removal of entropy from a temporal edifice that refines its possibilities.

What, then, is probability? If possibility is consequential, then probability is sequential. It is the unfolding of a temporal edifice that reveals its probable destiny to us. There is no trick in some 'so-called' predictions; they are at worst fake guesses and at best extrapolations from given evidence. If you see a car being driven erratically at high speed it is not a matter of clairvoyance, sixth-sense or supernatural powers to conclude that the driver may lose control of the vehicle, crash it and be injured and/or in legal trouble. But if the erratically driven car slows down and the driver appears to be acquiring some degree of control, then probability suggests that the vehicle will now be navigated safely. This is the point of Contingent Expansion; when a possibility evolves into a probability. As a sequence of eventuality progresses its contingent expansion becomes more definite, more set and more predictable. Probability takes expansion by degrees into the arena of Provisional Locking; it is the trend of causality.

Logic, the third element in determining the conscious understanding of Expansion Eventuality, builds upon evaluations of possibility and probability. It is not always concerned with prediction, because analysis of history also depends upon logic if it is to be valuable and accurate. However, logic is always concerned with removing ambiguity and subjectivity from the interpretation of information. More than any other intellectual discipline, logic aims to arrive at fact, regardless of how stark or injurious to the self. Logic has no room for pride, ambition, desire or self-service. It is the attempt of consciousness to touch the very fabric of causality.

Chance and causality illustrate the direct influence of natural law upon the progress of time; they are the widest possible scales of cosmic determinism but not opposites. They reveal, to any consciousness that may be observing, the realisation potential of specific temporal edifices in given circumstances. Logic proceeds from the inner coherence of chance and causality, appeals to objective reasoning and distils order from chaos.

Decision Eventuality

Having rejected the notions of prediction, predetermination and the 'bullet with your name on it' exactly how much room is there for consciousness to determine its own destiny? Surprisingly, in a universe with quite clear limits of possibility, finite creatures can have quite a substantial amount of influence but only in their own local arena. The event inertia of the cosmos dilutes down from infinity to the SNPWC scale, losing status and momentum as it filters through each successive stratum. In the mystic orient shaolin monks learn from childhood to control physical strength and thought to seemingly superhuman levels, yet these superhuman abilities, impressive as they are, effect only what is in proximity to the shaolin masters who have learned the skills.

As noted before, consciousness operates in the void between transmission and reception of information. The cosmos still ultimately influences the way in which information interacts with consciousness, across multiple waves and the QEA. We have choice, seemingly, and that choice can make an incredible difference, but we cannot override the unbeatable. Such as it is, our ability to influence the universe can vary between negligible to localised impressiveness. It is Decision Eventuality; the power to alter history by determined intervention.

But the more deterministic the cosmos becomes – and that means the more monumental the temporal edifice our sentience interacts with – the less choices are open to us, the less rein we have with the information from which our knowledge is developed. One way to look at the situation of consciousness is to recognise that cosmic information is a continuous wave of propositions. The alternative solutions with which consciousness responds to these propositions reflect the operating mechanism, status and efficiency of consciousness. As the cosmic status of determinism increases, decision eventuality reduces by inverse proportion; at one scale end there is the infinite macrocosm over which decision eventuality has no power. At the other scale end is the SNPWC, over which decision eventuality can wield its greatest potential influence.

The CSFC yields infinite information but consciousness intercepts only a finite amount; Decision Eventuality results from sentient predisposition, analytical accuracy and intervention colliding with history. It is a reaction choice in the realms of possibility but not always logic, and history shows that it may be good or evil, deliberate or accidental, organised or messy.

Quantum Weirdness in Expansion Theory

Quantum Weirdness is the blind spot of existence, the loss of information due to entropy. The reduction factor (rf) is a portion of unity (1), the equation for Quantum Orthodoxy being QO = Q - QW = Q x (1-rf) and for Quantum Weirdness QW = Q x rf. Broadcast efficiency (1-rf = β) is a unit figure in itself and represents the part of reality that is accessible to consciousness allowing for the degradation of Quantum Weirdness. Reception efficiency {R} is the potential maximum capacity of a unit of consciousness to acquire information. The Quantum Expansion Arc (Quexar) Q_E^A = Q - ØT and since ØT = rf then Q_E^A = Q - rf.

S = success rate of QEA = $\dfrac{Q_E^A}{Q}$ x 100

By substituting Q_E^A for Q in each calculation we obtain rough values;

Potential of transmitter = βQ_E^A and Potential of receiver = RQ_E^A

Para-eventuality = $_pQ$ = βQ_E^A x RQ_E^A

There is also the matter of the Quantum Interception Arc, very much a factor when dealing with conscious activity in universal transactions. Where; Q_i^a = Quantum Interception Arc or Quinarc, Q^1 = eventuality less minimum loss of clarity, q_0 = eventuality less maximum loss of clarity, A^0 = activation of observer in relation to temporal edifice, O^{sp} = Observer's Stimulus Potential = $\sqrt{A^0}$ and view-eventuality =

$_vQ$ = Q_i^a x O^{sp} x Q then Q_i^a = $\dfrac{Q^1 + q_0}{2}$ x A^0

Furthermore; Ω = Total information in a system, Δ = Pattern strength, μ = Comprehension of each unit of information, Ø = Total information lost; and T = number of time units, ØT = degradation so far = rf and
$\Omega\mu$ = total comprehension and, $\Delta = \Omega\mu - \phi T$ and $\Omega\mu = \Delta + \phi T$

$$\Omega = \frac{\Delta + \phi T}{\mu} \text{ and } \mu = \frac{\Delta + \phi T}{\Omega} \text{ so } \emptyset T = Q - \frac{Q^1 + q_0}{2} \text{ therefore } \Delta = \Omega\mu - \left(Q - \frac{Q^1 + q_0}{2}\right)$$

with rearrangement of terms $\Delta = \Omega\mu + \frac{Q^1 + q_0}{2} - Q$

Returning to an earlier illustration; $Q^1 = 0.9$ and $q_0 = 0.215$ then $A^0 = 0.88$ and

$O^{sp} = 0.938083152$ giving a result of $Q_i^a = \frac{0.9 + 0.215}{2} \times 0.88 = 0.4906$

and $_vQ = Q_i^a \times O^{sp} \times Q = 0.4906 \times 0.938083152 \times Q = 0.460223594 \times Q$

$T = 0.5$ (50%) elapsed seconds, entropy = 0.5 whilst $\beta = 0.8$ and R = 0.7
$\emptyset T = 0.2453$ and $Q_E^A = Q_i^a - \emptyset T = 0.2453$ whilst $\beta Q_E^A = 0.19624$, $RQ_E^A = 0.17171$
and $_pQ = 0.03369637$
Reducing from the original strength (1) to only 0.03369637 is a factor of 29.67678714 or

$$\frac{1}{\beta Q_E^A \times RQ_E^A} = \frac{1}{\beta R(Q_E^A)^2}$$

$Q - \frac{Q^1 + q_0}{2} = 0.4425$ so in the above example where $\emptyset T = Q - \frac{Q^1 + q_0}{2}$ and T = 0.5
the value for $\emptyset$ is 0.885 leaving only 0.115 of the original edifice integrity.

Since $\Delta = \Omega\mu - \left(Q - \frac{Q^1 + q_0}{2}\right)$ or $\Delta = \Omega\mu - \emptyset T$ then $\Delta = \Omega\mu - 0.4425$

The highest value that there can be for $\Omega\mu$ is 1, therefore the highest possible value

for Δ is 0.5575; exactly identical to $\frac{Q^1 + q_0}{2}$

All measurements are in relation to unity (1) and they are our most complete
calculations for quantum weirdness so far. Deterioration after conscious intervention is
clearly greater than Quantum Weirdness prior to consciousness. And the consideration of
conscious activity in the universe is not yet over........

Section G: Multiple Wave Conception

Perception, Preconception and Conception Eventuality

As its base unit this stage in our theoretical exploration of the incidence of consciousness with reality uses the Quantum Concept Arc or Quonsar - Q_C^A. Its two predecessors noted how information expands through time to meet units of sentience, a journey from transmission to reception that is fraught with pitfalls and reductions, and how sentience intercepts that information when it arrives. Unlike perception, preconception and conception do not require direct contact with information. The Quonsar measures sentient ability to correctly conceptualise the universe beyond, not the precise level of interaction.

Whilst it is true that our conception of existence, or preconception of it, might rely on information that has been received historically, the break with reality is a break with the present. Depending on various factors, that break may be slight or total. To derive necessary equations for this aspect of sentient interaction in cosmic affairs could prove rather more troublesome than I would like; it has been interesting to analyse the degree of reliability with which consciousness perceives the multitude of information stimulating it. Perhaps 'assaulting' it would be a more fitting adjective, because the CSFC releases a stream of information so voluminous that interaction is only possible with a minute portion; most information sweeps over our heads, unread, unchallenged and unnoticed. Even the most powerful intellects in the universe are going to miss the bulk of data passing them on its outward expansion through the CSFC.

It is clear that the relationship between a unit of sentience and the cosmos is a tenuous and unequal one at the best of times; at the beginning of Paper One I spoke about the 'Internal' and 'External'. Broadly these definitions illustrate the imbalance that is the main problem with consciousness; small units of limited ability attempting to comprehend infinity. By interpreting the reality it has perceived historically, a unit of consciousness derives concepts that amount to its outlook on life and can completely colour its reaction to future encounters with reality. This process of ongoing interaction between consciousness and reality is a complex process that filters through several stages before settling. We shall quickly examine this multistage filtration process to see how objective information about reality is translated into subjective outlook.

1. ELEMENTARY PERCEPTION
a. sensory faculties
b. sensory experience
c. knowledge

2. PERSONAL PRECONCEPTION
a. impression
b. assumption
c. prejudice

3. CONCEPTION
a. realisation
b. rationalisation
c. adaptability

Outlook is thus a grey area dominated by this triple triad of filtration, a deficiency in any one of which can seriously jeopardise the probability of a well-balanced, healthy and accurate concept of reality.

INADEQUACIES IN ELEMENTARY PERCEPTION

Defective sensory faculties can generate self-deceptions, erroneous data collection, lost or wasted information or even hallucination. Inadequate sensory experience of this type, particularly where it is recurring, leads to narrowness of vision. This in turn produces deficient knowledge resulting in poor visibility, haziness of comprehension and consequentially severe restrictions on progress and development.

INFLATED PERSONAL PRECONCEPTIONS

Overpowered impressions can cause outlook erosion by intentional or unintentional fraud of the self. An impression that is too strong and uncontrolled will drown objectivity. Overpowered assumptions pervert outlook by creating a wholly inappropriate matching of view to reality. Subjective and unwise prejudices divert outlook from the correct path of truth, thus compromising sanity and undermining objectivity.

UNDERFUNDED AND MISMANAGED CONCEPTION

A failure of realisation prevents the ripening and maturity of outlook. Incorrect or inactive rationalisations permit erroneous foundations in outlook to pass unnoticed and thus unchallenged. Unwillingness or inability to adapt result in an outlook that is brittle and unresponsive to new factors that could modify, develop or complement it.

In addition to the failures in these nine very vulnerable factors in the triple triad above, there is the clarity loss through Quantum Expansion Arcs and Quantum Interception Arcs to consider. Whatever has previously entered the psyche has already been processed and diminished by those arcs before it reaches the stage of absorption into the Quantum Conception Arc.

IN-WORLD and EX-WORLD

Again we return to the model of the Internal and External, as our exploration of the collision between consciousness and reality probes ever deeper into what is and what is not the case. With increasing complexity in the investigation of reality interacting with sentience, we need to develop the Internal/External model itself.

The Internal

This consists of the physical, objective, subjective and fictitious reality of a unit of consciousness. Many billions of such units exist, but ultimately they are disassociated from one another and so the situation of many billions of Internals does not in itself create greater accuracy in conceptualising the universe. The Internal can be divided into the 'IN-UNIT' (physical and objective) and the 'IN-WORLD' (subjective and fictitious).

The External

This represents all that is not the self, the sum of all temporal edifices that are beyond a unit of consciousness. It is the absolute reality that intelligence strives to know. The External can be divided into the 'EX-UNIT' (physical and objective) and the 'EX-WORLD' (subjective and fictitious).

In an infinite universe the EX-UNIT is infinite itself, causing us to face the uncomfortable but nonetheless true realisation that however well we conceptualise the EX-UNIT our outlook will be deficient. The problems of a sentient and finite Internal's ability to understand the infinite External in which it is immersed is magnified by the extraordinarily circuitous and multistage process by which infinite information is translated into finite blocks of comprehension.

It is hardly surprising that many Internals, even reasonably intelligent ones, abandon the sifting and analysing process having become disillusioned with their own progress. What is alarming, however, is that Internals can claim to be intelligent yet live out their lives content to end a process of engagement that most clearly defines the presence of intelligence. Worse still is the situation where Internals scorn the most perplexing questions of existence.

The Structure of Need & Want Psychology

Having established the dependence of our outlook on perception, preconception and conception it becomes clear that how reality gets filtered through the triple triad largely depends on personal motivations. And these, obviously, are a matter of basic animal 'need & want' to understand and control the temporal events and edifices encountered. It seems that to a certain extent we believe what we want to believe and ignore the rest, regardless of merit.

Consciousness at its most warped and insecure functioning tells itself that there is no reality beyond whatever it chooses to recognise. In its desire for false security, the weak mind believes that what it perceives is absolute reality. Subconsciously sentience tricks itself into the acceptance (as unequivocally true) of any information that has passed through the triple triad of perception, preconception and conception – regardless of how decimated it became at the end of the process or how incompetent the process is. History should have shown us how to correct this anomalous interaction with universal information, because history has taught us many important lessons. However, it seems that consciousness is not always able or even willing to learn from history.

Heracleides of Ponticus and Aristarchus of Samos both envisioned a heliocentric solar system, using logical deduction of the phases of the moon and the transition of constellations. They had no fancy technology to assist them in their deductions, only logic and a tireless desire for the truth, whatever that may be. Many of the ancient Greek philosophers arrived at surprisingly accurate deductions about the world and its place in the universe, through pure reasoning, mathematics and rationality. They saw not what suited them, not what fell in line with preconceived ideas or theologies; they looked for truth and accepted it no matter what it said. Geometry and trigonometry – twin temples of pure reasoning – were part of their method, but more than that an open-minded acceptance of evidence enabled their astounding successes. They needed and wanted cold, hard reality, and nothing else.

Following the unprecedented achievements of ancient Greek philosophy, what went wrong with the consciousness of humanity? For 1800 years the geocentric universe reigned supreme on Earth, bolstered and promoted by scheming and self-serving religious officialdom, until Copernicus committed the heresy of recognising the truth. Unfortunately science itself provides its own disturbing examples of sentient vulnerability to misconception. When Arhenius put forward his theory of electrolysis he was ridiculed. On the other hand the 'flat Earth' hypothesis was accepted for centuries.

What happens to the conscious view of reality when something is forgotten or we have become disorientated? Has reality really become fundamentally different merely because we are depressed, mentally disturbed or labour under a falsehood? If reality is controlled within the triple triad then its alterations would mirror the changes and shifts within the triple triad; imagination, moods and mental condition would create reality. It is my contention, however, that reality is something that exists before and beyond experience, sometimes becoming partially accessible through the filtration system of the triple triad, prior to assimilation by consciousness. The precise nature of reality is independent of witnesses or spectators, and the reality we know is therefore a dilute, inferior version of it.

The psychology of need & want is a way to understand our subjective intervention between reasoning and reality. It coalesces into a composite model;

A. **Survival.** Essential needs such as air, water and food. Essential aversions such as poisons or explosives.

B. **Hyper-survival.** Assumed needs such as money, noise or games. Assumed aversions such as competitor species or altruism.

C. **Comfort.** Essential wants such as warmth and friendship. Essential aversions such as cold or loneliness.

D. **Hyper- comfort.** Assumed wants such as chocolate trifle. Assumed aversions such as duty.

Although all our pursuits and actions fall into these four categories, there are only two major axioms that govern our interference between reasoning and reality.

Axiom 1; we act from motivation that we need or want to do something.
Axiom 2; we do not act when these motivations are absent.

At a superficial glance this may seem at odds with everyday experience; millions of people who possessed no death wish have placed themselves in mortal danger to protect other life. Surely then they acted this way even though motivation was absent? Not so; there may have been no motivation to face danger, but it was compensated for by the motivation to protect other life.

Levels of Reality

It has become obvious by now that consciousness must acknowledge its frailty, particularly as it strains across three different dimensions of reality to produce an accurate understanding of the unfolding universe. A failure to acknowledge this frailty is the greatest stumbling block that a mind has in its search for truth. This fits in with our 'need & want psychology' model; the survival of intelligent reasoning is critical for accurate comprehension and is therefore a necessity that is independent of our acknowledgement. It is an Ex-need and, like all Ex-needs, requires the process of acknowledgement in order to be realised and activated. Just as our need for food must be acknowledged by our consciousness in order to motivate us to eat, so our need for intelligent reasoning must be acknowledged in order to motivate us to interact meaningfully with reality.

This does not guarantee us from making errors, just as hunger does not guarantee that we will always eat precisely what we should when we should. But the acknowledgement and motivation of a need for food are starting points that are less destructive to us than their absence, which permits starvation. By analogy our search for truth will fail if we do not acknowledge the need for intelligent reasoning and become motivated through that acknowledgement. The greater the activation level is the closer we can approach reality. Again we are talking about bridging the uncomfortable gap between the External and Internal. And we are returned to our triple triad;

LEVEL 1: *Absolute (pure) reality* is external to sensory perception or the processes of reasoning, existing independently of efficient interception by consciousness. This is inducted into the triple triad through ELEMENTARY PERCEPTION (sensory faculties, sensory experience and knowledge).

LEVEL 2: *Subjective (face value) reality* is taken for granted, without awareness of vulnerability in sensory perception, inefficiency in reasoning or failure to counteract prejudicial or ignorant interpretations of phenomena. This is accepted erroneously through PERSONAL PRECONCEPTION (impression, assumption and prejudice). *Fictitious (imaginary) reality* is a suborder of this level. It is either *intentional* (contrived deliberately for a specific purpose) or *unintentional* (produced by faulty perception or the deceit of others).

LEVEL 3: *Objective (rationalised) reality* is acquired by sensory perception and abstract reasoning, subjected to intense scrutiny and balanced against available rational processes with a conscious desire to eradicate error. As the third type of reality this represents the higher functions of consciousness deliberately attempting to correct the failings at the first and second levels, through CONCEPTION (realisation, rationalisation and adaptability).

The universe exists in its current form as a result of a precise history, regardless of the extent to which consciousness is aware of it or the accuracy of comprehension. This is absolute reality (LEVEL 1). The idea that snakes are evil is standard subjective reality, whilst the events in Star Wars are intentional fictitious realities and someone believing an adulterous spouse to be honest is unintentional fictitious reality (all being examples of LEVEL 2). The Earth is the product of huge forces acting across 5 billion years, leading to an oblate spherical planet surrounded by a gaseous, breathable atmosphere with a relatively stable surface environment of land and water capable of supporting diverse life forms. This is objective reality (LEVEL 3).

The Acceptance of Quantum Eventuality

Taking these thought models further it is possible to see that reality is only related to the realms of perception, preconception and conception by a variable and often tenuous link, which can never be perfect. The first realm of our triple triad, elementary perception, is mainly inductive, being concerned with the primary absorption of information. The second realm, personal preconception, is mainly concerned with the emotional imprinting of personality upon that information. The third realm, conception, is mainly deductive and aims to improve the outcomes of the first and second realms using intellectual sobriety. To succeed at that noble goal it has to strive to sharpen the acuity of the senses, build a secure knowledge base and organise personal preconceptions into meaningful and appropriate contributions.

The acceptance of Quantum Eventuality largely depends on the second realm of personal preconception. This is the weakest part of the bridge between reality and consciousness for two reasons. Situated in the middle of the triple triad it carries the greatest tension between 'what is' and consciousness. It is the part of the triple triad most closely associated with entropy, and requiring the most effort to govern; anyone can have opinions but few are prepared for the hard work required to make their opinions obey observation (LEVEL 1) and thought (LEVEL 3). Since it carries the greatest weight in the daily operations of most sentience it is the one area most likely to buckle under its load. And it is the area most clearly involved in the Internal's conscious preoccupations with 'need & want' psychology.

This being the case, personal preconceptions divide significantly in line with the 'need & want' psychology model:

A. **Survival.** Essential needs leading to **Primary Acceptance Thresholds.** Essential aversions leading to **Primary Resistance Thresholds.**

B. **Hyper-survival.** Assumed needs leading to **Primary Acceptance Impulses.** Assumed aversions leading to **Primary Resistance Impulses.**

C. **Comfort.** Essential wants leading to **Secondary Acceptance Thresholds.** Essential aversions leading to **Secondary Resistance Thresholds.**

D. **Hyper- comfort.** Assumed wants leading to **Secondary Acceptance Impulses.** Assumed aversions leading to **Secondary Resistance Impulses.**

The Internal is at a disadvantage in its attempts to grasp the External. In objective reality (the nearest consciousness can get to absolute reality) the Internal has made a valid attempt to eradicate its disassociation from the External. In subjective reality the Internal has made little or no effort to overcome its disassociation, except in fictitious reality where effort has been expended to promote the disassociation; in some cases, such as art and music, this may lead to a surprisingly enriched outcome. Alternatively, consciousness may simply be incapable of preventing an enforced disassociation, as in madness.

Clearly the success of a unit of consciousness can only be measured in terms of how it relates to the External and whether or not its acceptance/resistance thresholds and impulses are accurately geared and its perception/reasoning adequate for their tasks. Evidently there are many variables in this psychological equation, and some unfortunately high error margins.

Attitudes, Impulses and Thresholds

The stronger these characteristics are, the more easily they are triggered. Most people have strong resistance impulses to pain, but a higher resistance threshold; it is quite normal to be capable of tolerating far more pain than one is prepared to. Only strange people seek out more pain than the body finds comfortable. Generally we can say that acceptance thresholds and impulses to particular experiences have correspondingly negative resistance thresholds and impulses to the opposite of that experience. But whilst resistance thresholds are higher than resistance impulses, it is usually the other way around for acceptance thresholds and impulses; our impulse to resist unpleasant experiences is less than our tolerance for them, whilst our impulse to accept pleasant experiences is greater than our capacity for them. $AI \geq AT$, whilst $RI \leq RT$

It is also typical to add the acceptance and resistance values together for any particular experience plus its opposite, and obtain zero. It is the psychological equivalent of the departure of pulsonic contours from their zero-base. The greater these A and R values are from the zero-base, the better defined is the character and the more prone to instability. Whilst $AI \geq AT$ and $RI \leq RT$ the consciousness can maintain general stability, but when the values reverse, unstable personality traits are likely to surface. There is also instability where the above aspects and their opposites do not cancel out to zero. Furthermore, acceptance thresholds and impulses should be naturally high for our essential and assumed needs and wants. Conversely, resistance thresholds and impulses should be naturally high for essential and assumed aversions. Again the risk of instability increases where these natural values are skewed.

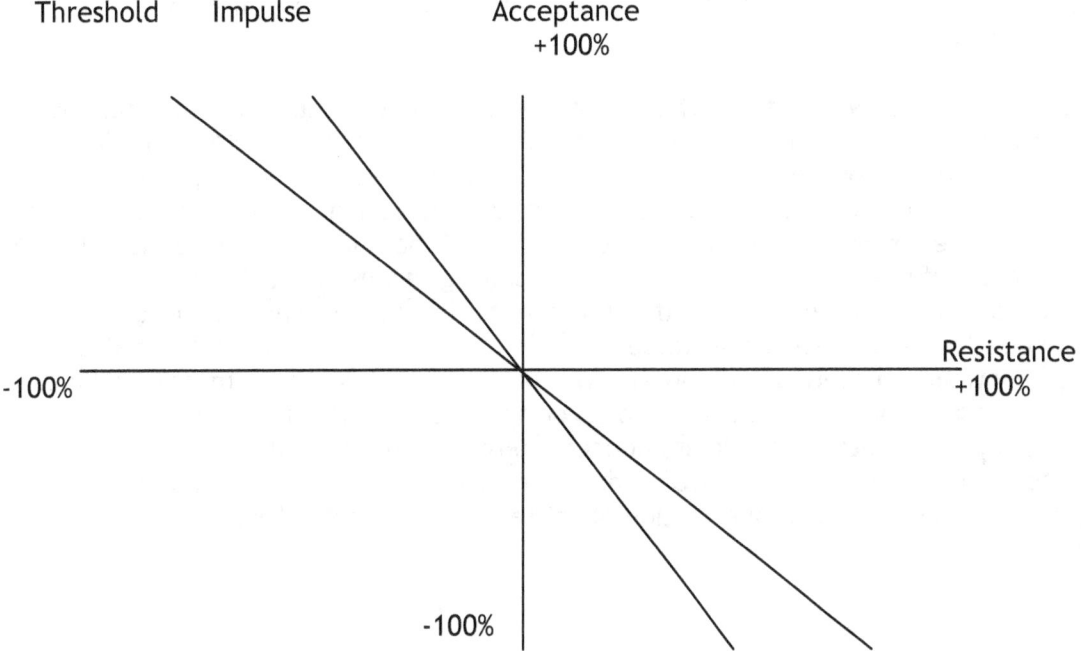

Fixed Ideas and Motive Entropy

Conscious minds acquire and use fixed ideas to help sift information and concepts in order to simplify comprehension of them. Fixed ideas can be developed at any stage in life, but those formed in childhood as a result of parental and other influences are the most ingrained and difficult to rationalise or modify. Only in very rare circumstances do fixed ideas undergo an examination rigorous enough to empower drastic revision, particularly improvement. Thus, in the majority of people, fixed ideas built up from the formative years onwards persist until termination. Since growth is a property of intelligence, and the lifetime retention of unchallenged fixed ideas represents a lack of growth, I submit that retaining unchallenged fixed ideas is not intelligent.

This is just one way in which humans fail to achieve their intellectual potential, and it is a prime indicator that the human species is not as intelligent as it likes to believe it is. There is an unfolding realisation that at every stage of the activation of sentience there is a frail dividing line between success and failure in its link with reality. Exactly which side of the roof the raindrop will fall is uncertain, but which way it is pushed is another matter. We have seen how attitude – the sum of acceptance and resistance impulses and thresholds – affects the preconception stage of reality, tinting it with subjective rather than objective overtones. But the impression, assumption and prejudice that govern personal preconception are themselves the apprentices of a deeper, more sinister and entrenched master.

Motive. The psychological basis for all actions is self-aggrandisement. This is true of any conscious mind, whatever its level of intelligence and whatever species it belongs to. Self-aggrandisement may take many forms, bad or good. Sentience might pursue the acquisition of unnecessary property, fame, vice, the climbing of political or corporate ladders, the domination of other people's thoughts and emotions and of course power generally over others. At their worst these motives can reveal the worst in humanity; the desire to depose others from their comforts and survival, or to force them to suffer in our place. Greed, hatred, cowardice and ignorance may abound at this end of motivation, and entropy is here at its most alarming strength, consuming the sentient being concerned.

But all is not exactly wonderful at the best end of motivation. Such laudable goals as efficiency at work, experience of harmless entertainment, bravery, helping the less fortunate, bestowing our love on others, humanitarianism, vegetarianism, selflessness, purity and education appear to be beyond reproach or question. But unless we are very careful and rare beings indeed, even the best of us will be pursuing these aims - at the deepest levels of psychological motivation – for self-aggrandisement. We want to be admired, loved, respected, and proud of our own achievements. These aims are certainly better than those at the worse end of motivation, but they still have at their centre how we see ourselves and how we wish others to see us. And it must be admitted that any being having made a decision to do or not do something, no matter how altruistic their motivation, of primary concern to them is the role of their own desire and perception. In short every unit of consciousness is a slave to their own quantum conception arc and whatever motivation of self-aggrandisement filters out of their triple triad.

Since motive, even at its purest filtration, detracts from reality and objectivity it represents a manifestation of entropy; hence motive entropy – the entropy associated with the will of consciousness.

Calculating the Variety of Consciousness

The calculation of Quantum Concept Arcs (Quonsars - Q_C^A) is a result of the filtration process in the triple triad. Again every value is expressed as a portion of unity, so a sensory faculty value of 50% is 0.5 with 1 being the ideal standard of zero entropy. With this in mind the following formula presents itself;

$$S_f \times S_e \times K = E_p$$
$$(i-a)x(1-pj) = P_p$$
$$R_e x R_a x A = C$$

$$Q_C^A = E_p \times P_p \times C$$

Definitions of terms;

S_f = sensory faculty strength

S_e = sensory experience strength

K = knowledge strength

E_p = total strength of elementary perception

$(i-a)$ = impression less assumption

$(1-pj)$ = 1 less prejudice

P_p = Personal preconception final strength

R_e = strength of realisation

R_a = strength of rationalisation

A = strength of adaptability

C = final strength of conception

Obviously there is great scope for deterioration, even if the values of each contributing factor are high to begin with. Only consistent values of 1 lead to a final value of 1, but anything less than one spells trouble. Even impossibly high consistent values of 0.99 lead to a final value for Q_C^A of only 0.922744694; and dealing with more realistic (lower) values has even more worrying results. Suppose there are across the board valuations of 0.5; we end up with a Quantum Concept Arc of only 0.00390625 so that things are not looking good for consciousness. We should look on the bright side, though; a conception value of 0.00390625 of the photo-cosmos still gives a supreme scope for our comprehension, providing that we remember to conceptualise as much of the entire photo-cosmos as possible. If we only conceptualise our own back yard then this figure means we **_will_** end up drowning in our own ignorance.

With the Quantum Interception Arc the final reality value was $_vQ = Q_i^a \times O^{sp} \times Q$ and was termed view-eventuality. Then Quantum Expansion Arc had para-eventuality with a value of $_pQ = \beta Q_E^A \times RQ_E^A$ so now we require the parallel for Quantum Conception Arc; Quasi-eventuality, $_qQ$ with its factors $_qT \, _qV \, _qM$ is derived from the simple relationship

$$_qQ = \,_vQ \, _pQ$$

Only one more task remains in this Quantum Conception Arc theory; to categorise different types of consciousness according to their creative success in interacting with reality. It is a bit like a maximum activity potential for conceptuality.

Given that the brain has a processing function with regard to information, what exactly is intelligence? Is it a measure of how high the absorption of information is? Is it a measure of how successfully absorbed information is processed and comprehended? Or is it a talisman of the retention of information and the appropriateness of the brain's reaction to it?

Firstly we need to dispel some obvious myths. A computer with a 100 Gb hard drive can contain far more information than one with only a 5 Gb hard drive. Furthermore, the computer with the smaller drive may be filled to bursting, say up to 4.97 Gb, at which point its smooth functioning begins to be impaired. On the other hand the computer with the 100 Gb drive may only have 2.1 Gb filled; all that extra capacity is simply not used, and its actual possession of knowledge is less than the 'weaker' computer. To be sure, nobody could call the one computer more intelligent than the other, whether making a comparison between their potential (disk space) or retention (disk usage).

So, ability to possess information – the raw capacity to hold it – and information retention – memory - are two qualities that are not necessarily evidence of intelligence. Some extremely intelligent people can have poor memories whilst average souls can display astounding feats of memory. We see this in the astonishing amount of information computers can hold and maintain, compared to us; we do not assume that this makes those computers more intelligent than us. And computers retain their knowledge when turned off, as well as their ability to reactivate; we do not. Once turned off a human's knowledge is forever lost. Moreover, books require no power to hold and retain their information.

Now we approach the crux of what intelligence really is; the book and computer have no reaction to the information they contain. They have no comprehension or discretion concerning the nature, validity or moral purpose of their contents. Their zero reactivity value towards information is a straightforward reflection of their inanimate essence. So it is hardly surprising that when one sees sentient beings displaying no reaction to stimuli a question arises as to their possession of life and functionality. Even plants react to stimuli; it is a sign of life. The more complex and evolved a sentient creature is the more notable and diverse its range of reactions should be: the more it should react to subtle and intricate stimuli. That leaves the matter of nature. If intelligence is best demonstrated by some sort of reaction to a stimulus it remains only to observe the nature of the stimulus and how it has been reacted to (that is how appropriately). It follows that negative stimuli should be rejected and positive stimuli should be embraced, but that rarely happens.

Thus the following intermediate categories present themselves;

<u>Stimulus Negative</u> - biased in favour of the negative. Range 0.01 to 0.99 $_f Q$; includes the subsets Minimal Gatherer and Irreverent Gatherer.

<u>Stimulus Neutral</u> - there are no responses. Range unity (1) $_f Q$; - includes subsets Inanimate and Neutral Gatherer (unaware).

<u>Stimulus Positive</u> - biased in favour of the positive. Range 1.01 to 1.99 $_f Q$; includes subsets Active Gatherer and Generator Mind.

This information shows briefly how intellectual potential can be wasted, so that someone with a powerful mind that is malfunctioning can end up with a lower score than a less powerful mind in full working order. The average citizen of a fully civilised and intelligent species should score approximately 1.25 to 1.3, whilst a lowest scoring individual for such a species would be in the region of unity. The human race scores an average of 0.5, which is well below minimum for a truly civilised and intelligent species. At this juncture in time the human race clearly has no right to consider itself either intelligent or civilised.

Section H: Selective Wave Transmission

Principles of Linguistic Eventuality

The basic unit of this theory is the Quantum Language Arc or Qualar - Q_L^A and it measures the efficiency of language to convey meaning. Language as a means of communication is a restricted medium founded on a selected and arbitrary group of bases. The bases are set out in the following table.

Symbol	Definition	Function	examples
N	Object noun	Nominative	*Time* waits for nobody
n	Subject noun	Nominative	She thought about *time*
P	Pronoun	Nominative personal	*I* saw *you* yesterday
T	Truth	Clarity	
F	Falsehood	Obscurity	
?	Interrogative	Uncertainty resolution	Why, what, where, when, which, how
±	Possibility	Alternatives	Maybe, sometimes, perhaps, or, instead of, other than, different from
X	Exponential	Generation	More than, greater, great, beyond, most
÷	Entropic	Reductive	Less than, fractured, small, within, least, lesser
+	Positive	Incremental	And, with, as well as, plus, also, yes, in addition to, at
-	Negative	Destructive	Without, less, minus, not, no, neither
$_fQ$	Focus eventuality	Quantum definition	Description
$_fT$	Focus time	Linguistic time	Clock measurements
$_fV$	Focus volume	Linguistic volume	Dimension measurements
$_fM$	Focus mass	Linguistic mass	Weight / force measures
Δ	Verb	Action	To *construct*
∧	Modifier	Qualification	Prefixes, suffixes, articles, comparatives, superlatives, adverbs, adjectives
=	Equivalence	Passage	Same, equal to, through, by, across, over, during, become
∂	Causative	Decision eventuality	Necessity, therefore, because of,

Examples of how these bases are arranged into coherent structures:

The train arrived at the station late.

∧NΔ∧n∧

Having walked for ten years, Bill bought a car.

$\Delta_1 \, {}_fT \, N\Delta_2 \wedge n$

People change over the years.
$N\Delta = \wedge n$
Most people are the same at a basic level.
$XN\Delta\wedge = +\wedge n$

Sodium chemically combined with chlorine becomes table salt.
$N\wedge\Delta + n_1 = n_2$

Periodically comets approach the sun.
$\wedge N\Delta\wedge n$

Geese fly whilst the summer lasts.
$N\Delta_1 = \wedge n\Delta_2$

Explosions cause a mass to disassociate and expand through space. $N\Delta_1 \wedge n_1 \Delta_2 + \Delta_3 = n_2$

$5x + 3a - 4 = 0$
$N + n_1 - n_2 = n_3$

$2a^2 - x^3 + y = 0$
$N - n_1 + n_2 = n_3$

From the above it is possible to draw some interesting conclusions. The conversion of language into symbols repeatedly presents the pattern $N\Delta\wedge n$, sometimes in a different order (linguistically speaking, syntax) and with or without various qualifiers or modifiers. But essentially language attempts to describe the universe as a collection of actions being executed by one thing against another. This might not seem like much, but it is a staggering claim for language to make and immediately throws up the following inadequacies;

1. The only way language can describe an infinite number of events is by using an infinite number of $N\Delta\wedge n$ base groups, which is clearly not possible. Language is therefore unequal to the task of conveying complex universal information or anything beyond the descriptive capacity of the base group $N\Delta\wedge n$.
2. This shows that language cannot adequately describe situations in which actions are not being executed by one thing against another - and can therefore not adequately describe static relationships or conditions - unless mathematical equations are introduced.
3. The qualifiers and modifiers are fairly responsible for the ambiguity of language, and hence many of its inadequacies.
4. Every statement made in language without adequate reference to (or logical deduction from) relationship has questionable validity.
5. Every statement therefore requires to be followed with an unvoiced interrogative T/F?
6. Languages deal with specifics that are questionable, mathematics deals with abstracts that are provable. Mathematics therefore presents a clearer path to the truth.

There are two basic constructions in language; Interrogation and response. Every question seeks an answer and every statement is an answer, whether there was a question or not. Even a single word or name spoken in isolation is a contracted statement of something and thus an answer to an unspoken question. It is the precise nature of statement and answer construction that defines a language. We have already seen the general pattern of statement construction: $N\Delta\wedge n$, with an unvoiced interrogative T/F? Mathematical descriptions have the structure $N\wedge n_1\wedge n_2\wedge n_3,....$ They have easier to control outcomes with more obvious flaws or strengths mainly because meanings and definitions are condensed, require no bulky modifiers or specific facts and incorporate a testable conclusion automatically.

Consider the following linguistic structures;
How can Henry say I owe him money? :::: Who wrote the Eroica Symphony?
$?\Delta_1 N\Delta_2 P_1\Delta_3 P_2 n_1$ $?\Delta\wedge N$

Both of these sentences are questions containing the base structure $?\Delta N$. The answers to them are **Responsive Statements**;

Because he leant you money and you have not returned it.
$\partial P_1\Delta_1 P_2 N{+}P_2\Delta_2{-}\Delta_3 P_3$

Beethoven (wrote the Eroica Symphony).
$N\ (\Delta\wedge n_1\ )$

Either of these answers could just as easily have been announced independently of a question (with the removal of 'because' from the first sentence) and so stand just as effectively as **Creative Statements**. And neither may be true; the information they contain could be incorrect by design or ignorance.

Consider the following.

Tchaikovsky composed six symphonies.
$N\Delta 6n$
How many did he compose?
$?{+}\Delta_1 P\Delta_2$
Six.
6

These three sentences could have been more succinctly put if we think along the lines indicated earlier: $N\Delta\wedge n$ T/F? thus;

Tchaikovsky composed six symphonies. (Is this true?)

The interrogative add-on is the unvoiced and mostly ignored factor of any statement in any language, but – as said before – even a **Creative Statement** is a response to an unspoken question. The statement *'Tchaikovsky composed six symphonies (Is this true?)'* is a response to the unspoken question *'How many symphonies did Tchaikovsky compose?'*

The 'T/F?' add-on helps us to remember to question the assumptions of language, in a manner worthy of Socrates. It makes us investigate the truth of what we say, and so with respect to our assumptions about Tchaikovsky's symphonic output actual investigation leads to a different number from that previously indicated. The official line is six, but this rather blindly ignores the posthumously reconstructed Symphony No7 in Eb major, and the highly programmatic 'Manfred' symphony written between his No4 and No5. I remember sitting a music test at school in which I answered this very question with the true '8' and was failed on the question by the music teacher. I took this up with her after the test, and she stubbornly adhered to the official '6' put forward by textbooks, critics and musicologists. She challenged me to prove my answer and I did, by taking to school my vinyl LPs of Manfred and No7. She still refused to revise the score on my test paper; adherence to an unchallenged falsehood, in the face of a challenge.

Semantic and Linguistic Levels

As if the previous three arcs – interception, expansion and conception – were not sufficient detractors from reality, clearly language is the fourth and final straw. It may be a matter of conjecture to decide exactly how far language falls short of adequately describing reality, but I am willing to hazard a guess of 70% maximum capability. By that I mean that a well-constructed language could potentially only convey 70% of the reality it attempts to reflect. The true figure is likely to be even less than this, because language contains only finite amounts of linguistic quanta and thus only finite permutations. The universe of reality is infinite. So this is where the concept of focus eventuality ($_fQ$) with its components $_fT$ $_fV$ $_fM$ arises. It is the product of $_qQ$ multiplied by the % functionality of language.

And then there is the individual's grasp of language contributing to the degradation. Poor vocabulary, syntax, grammar, punctuation and comprehension will escalate the demise of reality into the pale approximation that is the Quantum Language Arc. It does not require a mathematician to see that someone with only a 50% grasp of a language that is only 70% efficient risks carrying only a 35% chance of understanding anything fully. This is an additional degradation, it must be emphasised, and is easy enough to illustrate.

Consider the sentence 'she was 150 cm tall'. Exactly how much information is this accurately carrying? It might seem a lot but is it? The fact is that only one portion of the sentence is provably accurate in a stand-alone way; the mathematical value 150. The rest of the sentence is almost entirely non-provable and thus merely speculative. The precise value of 1cm is not even hinted at and particularly there is no specific agreement on the valuation of 1cm between the transmitter and receivers of this information. The important matter of whether everyone can agree precisely on what they understand to be a 'cm' is therefore not satisfactorily addressed. Furthermore there is absolutely no information regarding the accuracy of the method used to measure 'her' height; if a tape measure was used can its precision be assured, or the competency with which the person carrying out the measurement actually read it? Were the results verified independently, was 'she' standing level or on tiptoe, is the measurement from toe to head or toe to up-stretched arms? And the tense, 'was', is also important; she may have once been 150 cm tall but is she still that height and when precisely in time was she 150 cm tall?

Clearly there are enormous problems engaged in just a few words intended to convey a single quantity of simple, apparently straightforward information. Only the numerical value can be held as beyond question; whatever inaccuracies or doubts may be inherent, her height can always be divided into 150 segments, equal or unequal.

Natural and Phenomenological Levels

There are two main stages in the construction of natural and phenomenological communication. They relate to the dual elements of language noted earlier; questions and answers or interrogation and response. In order to widen our concept of linguistic application I am renaming these two elements **call and response**.

Every temporal edifice and every event within and surrounding that edifice is a call; it is the universe asking a question. How the call is reacted to is the response. Now, where there is no consciousness (and hence, no intelligence) present, matter can be forgiven for failing to respond to the events, structures and fundamental principles of the calls generated by the universe. But where consciousness exists - especially consciousness that considers itself intelligent - the failure to respond is unbelievable. There is no point having the ability to contemplate the universe but remaining inert and no right to be inert but claiming otherwise.

Even without language, every call output by the universe is coded; consciousness receiving the call needs facility in that code. Facility is acquired through language. The concept of **call and response** bridges the gap between phenomena and description; parity between reality and linguistics. We need not pat ourselves on the back for having language; when one part of the universe creates a call it may be responded to unconsciously by another part that is not even alive. A meteor straying into a planet's atmosphere and gravity well creates calls to which the atmospheric and gravitational reactions are responses. Language is only a tool used by consciousness in order to cope with calls. And it is vitally important that the full principles and capacity of language are utilised if it is to be a helpful tool.

Learning through Language

The learning process requires three levels reminiscent of the triple triad in Quantum Conception Arc Theory. The main difference is that true learning cannot proceed unless harmful parts of the triple triad are controlled. To learn we must distil the important elements from the Quantum Conception Arc, discarding the rest. We end up with just three factors;

1. Induction (questions/calls)
2. Construction (unverified responses)
3. Deduction (verified responses)

It follows that learning begins with induction, a process desired by intelligence because it seeks answers. In the absence of a healthy induction process, and particularly with the handicap of poor deduction, construction creates warped caricatures of reality that aid ignorance; assumption and prejudice. Powerful deductive capacity is in many ways the most important of the three factors, because it alone can extract valid conclusions from any amount of evidence, no matter how small. Deduction shows how intelligently intelligence is applied. With both a healthy inductive and deductive capacity a unit of consciousness can afford the luxury of a healthy constructive counterbalance; this is the seed of creative genius – the generator mind - but it is not always realised. How well consciousness employs a chosen language reveals its dexterity in communicating with a body of information, but it may not communicate with other units of consciousness.

Levels of Actuality

1. **Absolute**; the universe of infinite calls. Quantum Eventuality that actually happens.
2. **Personal Induction/Deduction**; the in-world of objective responses. View-eventuality and para-eventuality – what we see happen.
3. **Personal Construction**; the in-world of subjective responses. Quasi-eventuality and focus eventuality. What we say happens, both to ourselves and others.

Once more the triad configuration is maintained. To appreciate the strength or weakness between absolute reality and personal construction we need only consider the psychological problems of the human mind. Like all phenomena, these have their own expression in language. But they may not merely acquire a linguistic expression; they may owe their root cause to the catastrophic mismanagement of language. The pressures and images that lead to mental dysfunction and eventual collapse possess a linguistic counterpoint that could attack the mind more directly and more devastatingly than the pressures and images alone.

Consider the office worker who suffers a breakdown as a result of work pressure. Is mental and emotional collapse really the result of demands and deadlines, or is it the result of the linguistic attachments and definitions assigned to such matters within the office worker's mind? The key element in either improving or eroding the tenuous link between absolute reality and personal construction is personal induction/deduction. Between what actually happens and what we say happens, what we see happening is regulated by logic and structure.

Without the powerful processing algorithms that logic and education provide, the mind is pretty much undefended against the degradations of entropy. And it is not sufficient to have merely one but not the other; a logical mind without sufficient study may well feel cast adrift in a sea of information that overwhelms it, whilst an educated mind without logic lacks the sound framework needed to efficiently manage its knowledge.

In that extremely important middle ground of personal induction/deduction any weakness can potentially lead to the following failures;

A. Incorrect identification of object and subject nouns and their definitions and modifiers. Example; believing that losing a game or contest is 'tragic'.
B. Incorrect application of causative or equivalence relationships. Example; believing that inanimate objects possess malicious intentions.
C. Overemphasising importance of pronouns 'I' or 'we' whilst undervaluing importance of other pronouns. Example; We deserve, they do not.

D. Incorrect application and usage of exponential, entropic, positive and negative qualifiers. Example; I cannot do any more.
E. Misapplication of verbs. Example; The human race is civilised.
F. Absence of interrogatives, leading to confusion between clarity and obscurity. Example; Everyone is equal.

A reasonable application of thought reveals the potential disaster in these failures, but absence of interrogatives is by far the worst offender. The failure to desire appropriate responses to the calls of the universe demonstrates deep psychological dysfunction that is far worse than any failure to develop appropriate responses. The mind has many limitations concerning how much, and precisely what, information it can adequately process. But if it chooses to abandon applying interrogatives to the processing of that information then it has already lost its battle with entropy; instead it becomes unable to distinguish between fact and fiction, reality and fantasy. In such a perilous relationship with information, truth can become lost and fallacy accepted. In order to avoid this condition, healthy minds desire true responses; their desire to know more may not always bring the answers sought, but it ensures that some standard is applied to the sifting process. And it is far more likely to bring some worthwhile answers than a desire to remain ignorant.

Evolution of Linguistic Comprehension

Tracing this process is easy enough; it starts in childhood with the acquisition of technical fluidity and the construction of a knowledge reservoir. Obviously progress will be according to the nature of the infant mind; <u>Stimulus Positive</u> will forge ahead faster and more surely than <u>Stimulus Neutral</u> and even further behind will be <u>Stimulus Negative</u>. It is not by coincidence that those least adept with language are likeliest to demonstrate an attitude of casual dismissal. And if they are unwilling to develop linguistic competence then their progress in any area of learning will be stunted.

Having acquired technical fluidity and a sound reservoir of knowledge, the young mind will quickly progress to artistic application; the imaginative dimension of building images with language. It is the beginning of creativity in any field, and most noticeably accompanies childhood in the form of story telling.

From this comes the leap to universal recognition – identifying patterns and links in knowledge, and understanding their cosmic significance. This is an accomplishment of maturity in language and it is a delicate stage, which can easily be lost if an individual is not careful. Looking at the human race, it appears to be the stage that most easily fails; progress without it is slight and in danger of becoming twisted into a mockery of intelligence. Failure at this level is also responsible for many personality disorders, most human bad behaviour and the proliferation of bad attitudes and short-sighted actions.

Philosophical judgement proceeds naturally from universal recognition, so the failure in one is failure in the other. Philosophical judgement is the acceptance of truth as universal law for self-control. It is the flourishing of good character and an honourable way of life and without it the final two elements of linguistic evolution are virtually impossible; spiritual resonance (awareness of one's responsibility to harmonise with nature) and the cosmic mode (directing one's own life toward total objective participation in existence).

The Specific Interpretation of Meaning

Earlier on I pointed out that every statement has a silent question, so that N∆∧n T/F? is the appropriate underlying structure. The interrogative automatically questions the validity of every statement and stimulates us to pursue the truth, rather than accepting information at face value. This may seem the province of philosophy rather than linguistics, but any meaningful application of language begins with the desire to seek accuracy, a core aim of philosophical thinking.

Consider the advertising paradox.

Statements from advertiser; 'everyone buys this product'. 'More and more people are buying this product all the time'.
Purpose of advertising; to encourage more people to buy a product.
Conclusion; the advertiser is lying to us. In their view not enough people are buying the product and demand may even be reducing.

Consider the advertising claim.

'This product is the best quality and value for money and is extremely popular'. Advertising reveals an inadequacy in sales revenue; the product is not selling as well as the advertiser would like. Perhaps the demand is lower than expectations because consumers question the value or quality of the product; if it needs to be promoted then it is not extremely popular. The advertiser is lying to us. Similar untruths and deceptions exist alongside the iniquitous world of advertising, in big business, the media and global corporations and government. Partners in crime and stupidity, it pays the greed of such powerful organisations to brainwash whole populations into believing convenient and transparent lies.

For example, that someone who moves a ball around for a living is an interesting super-being, or that a musically illiterate pop star is talented. It is a scurrilous thing to lead the unknowing to the abyss through corporate deception, then push them over with their own credulity. It is scurrilous because the unknowing will readily accept whatever they are fed, especially when everyone else gorges on the same delusion. It pays because, having convinced a gullible majority that ball-movers or rock stars are 'brilliant', industry can feed off the profits for ages.

There are rich pickings in ignorance so the independent, educated citizen is the worst enemy of governments and huge corporations. No amount of coercion or hype will ever convince a thinking mind that someone who moves balls around is of major importance, or that someone making an ugly noise on an electric guitar or drum kit is 'musically gifted'. Or that any of the products, policies or opinions such people endorse are worthy of support.

Governments and profit-greedy businesses rely on the mass hysteria principle; convince enough people and others will sign up, so long as the belief remains unchallenged by a moment's careful thought. The population is swept unthinkingly towards whatever is easiest, just as flood- water takes the easiest route, or stray matter falls helplessly into a black hole, picking up more mass as it travels. Eventually the situation is out of control and difficult to correct. Governments and corporations have the population right where they want them; drugged, duped and vulnerable to majority suggestions. Attempt to show anyone that the person making that horrible noise is a charlatan or that moving a ball around is not special and you risk anger, invective or even violence.

If 'everyone does x' or 'x is fantastic' are untrue then are their counter-statements true?

Everyone does not do x
X is not fantastic

It is possible that some people do x, therefore it is inaccurate to make both the statement and its counter-statement. There is not one person who knows everyone else sufficiently well to be aware of all that they do or don't do.

It is possible that x is of a quality between fantastic and not fantastic, that it is fantastic in some ways but poor in others, or is fantastic to some people but poor to others. Without the input of educated appraisal the business of identifying value in anything is hit-and-miss. Take music, for example. To someone deeply educated in music, and consequently refined in musical aesthetics, anything lacking in melodic strength, structural integrity and harmonic imagination will be meaningless noise, possibly even offensive. To anyone without musical education or developed musical aesthetics, such sound will be somewhere between acceptable and terrific. To the musical mind the work of such an intellect as Beethoven will immediately impress and satisfy; to the unmusical mind Beethoven will pass by unregistered.

A further property of meaning is equivalence.

If x = y then by asking what x is we are also asking what y is. Defining event-objects or temporal edifices with linguistic sounds does not seem to get us very far. If we understand by a word 'x' the representation of a reality 'y', then our understanding of 'y' depends heavily on our understanding of 'x'. Our earlier example, concerning the number of symphonies written by Tchaikovsky, does not help our understanding of the situation if we do not know what a symphony is, who Tchaikovsky was and do not understand numbers. Even if we do understand the realities represented by 'Tchaikovsky', 'symphony' and 'eight', hearing the information rendered in an unknown language will probably not register. Language, then, is no more than a string of synonyms for reality, and synonyms are sometimes chosen incorrectly.

Equivalence is even more painfully frail if we consider a string of connected linguistic synonyms. Here the capital letters represent absolute reality and small letters represent our grasp of it.

$$A \rightarrow B \rightarrow C \rightarrow D \rightarrow$$
$$\downarrow \quad \downarrow \quad \downarrow \quad \downarrow$$
$$a \rightarrow b \rightarrow c \rightarrow d \rightarrow$$

Now we have the added complications of $a \rightarrow b$ only being true if $A \rightarrow B$, $A \rightarrow a$ and $B \rightarrow b$ are all true. And this is just the situation between two factors, A and B. With absolute reality there are an infinite number of 'bits' of information available and thus an infinite number of opportunities for failure between the universe's transmission of information and our linguistic reception of it. And language is vitally important in the efficient running of human consciousness.

The Specific Interpretation of Intent

In our communications with others we may have two basic intentions; to present the truth or to present a corrupted version of it. In the first situation our main battle is with ourselves.

Telling the truth.

Even if we have the desire to communicate accurately, there are two problems; our efficiency and our intention. There is a limit to how far we can maximise efficiency or minimise deficiency, but our influence over our own intentions is considerable. We may tell someone truths in a way that is intended to hurt them, and that is wrong. Or we may protect them from a truth that would destroy them if revealed. The role of language in the promotions of honesty and intention is crucial; without language there is little possibility of deception and none of emotional cruelty.

Corrupted truth.

Whenever words are bent to our will, there is the possibility of their misuse. For those who do not care about their own literacy there is the danger that they may end up caught out by their indifference to language; they are at the mercy of the smooth talkers, confidence tricksters and liars of the criminal fraternity and big business, even the media. All forms of comprehension in all subjects and for all units of consciousness begin with a thorough command of language. Remove that and the ability to achieve awareness in each situation is diminished. It is not merely the failure to cope with grammar, syntax and punctuation; vocabulary is vital. To function in a linguistic society a unit of consciousness must understand a comprehensive vocabulary.

The understanding of each and every subject begins with words; understand the words and the concepts have a chance of falling into place. It is not possible to understand an academic discipline without the vocabulary in place. And it is not possible to counteract the convenient lies of governments, large corporations and the self-serving media industries if the false word constructions they generate cannot be seen through. For example, if government scientists claim that 'something' is safe, in ignorance we might believe them. But if we know the vocabulary of the science concerned, and therefore the building blocks of its concepts, then we can judge for ourselves whether 'something' is in fact safe.

Like Charges Repel

In physics, particles with the same charge type repel each other, whilst particles with opposite charges attract each other. In the meeting of linguistics and reality, a statement of negation belongs with a negative fact and a statement of affirmation belongs with a positive fact. For example, STATEMENT "I did tell them to wait" belongs with the FACT "I told them to wait". However, the STATEMENT "my skin is blue" does not belong with the FACT "my skin is not blue". Lies, deceptions and inaccuracy are the results of any mismatch. How can we resolve the discrepancy between particle physics and language?

We can imagine that the reason two similarly charged particles are repelled by one another is a result of the saturation of charge fundamentals in the SNPWC. In section D the investigation of the Quantum Pulse Cascade delivered the following definition for particle charge:

The main charge fundamental is the particle charge component (it collaborates with the particle mass component to create gravity) and it is measured in increments of a third from -2 to +2. Thus the pcc fundamental has thirteen possible values (-2, -1⅔, -1⅓, -1, -⅔, -⅓, 0, ⅓, ⅔, 1, 1⅓, 1⅔ and 2).

Charges try to balance around zero, for stability. When nature forces a particle to hold a particular charge it creates a slight instability in the main charge fundamental for that particle. Two particles with negative or two particles with positive charges together causes even further stresses in the stability of both, forcing them apart. They reflect each other's like charges back by mutual repulsion. Two particles of opposite charge accept each other's stabilising, counteracting charges and thus are drawn together.

In language a negative statement belongs with a negative fact because of a different type of reflection; not of forces but of truth. Similarly, positive statements belong with positive facts.

Entropy, Caution and Communication

The loss of clarity from fact to language is unavoidable and is a major factor in the entropy of communication. Although we cannot eradicate this degradation, we can minimise it. Clarity of language and the eradication of superfluous words, particularly of a subjective, descriptive nature, are good starting points. Saying what we mean, using correct grammar, vocabulary and punctuation can also reduce confusion. As seen before, however, the intent of the speaker is of crucial significance. Honesty and integrity on the part of units of consciousness transmitting information is vital to reduce entropy. And on the parts of recipients the experience to interpret the honesty content of what they receive and the wherewithal to react appropriately are vital.

The only unquestionable honesty in universal information transmission is between two temporal edifices that are not conscious; only then are the motives of either transmitter or recipient above reproach or question.

In saying what we mean and what is as near to the truth as possible, and as far from any personal distortion of truth as possible, our greatest ally is caution. Think before speaking. Consider the following montage.

"James Bond knows more secrets than most people." $N\Delta XnXn_1$ T/F?

Remember our basic linguistic structure $N\Delta \wedge n$ T/F?

But what does this sentence actually mean? Statistically speaking, the sentence means that a much higher percentage of Mr Bond's knowledge consists of information unknown by most other people. There is an implication that we are talking about confidential government information that is deliberately withheld from the populace. However, specific details of the information that is being referred to is not revealed in the sentence, therefore the assumption is dubious and based hypothetically on our further assumption that the James Bond being referred to is the fictitious spy of the celluloid variety. The sentence has further lead us to these assumptions by use of the word secret.

Suppose that in actual fact I am talking about a car designer called James Bond who is currently working on a revolutionary car design, or an author writing a controversial expose of the chemical industry. What happens if we modify our sentence?

"James Bond knows few secrets."

Does this mean there are only a few bits of information that Mr Bond does not have access to, or that there are only a few things he knows that are not widely known?

The terminology is vague and this is further enhanced by the inherently incorrect juxtaposition of 'know' and 'secrets'. If you know something it cannot be a secret to you; the fact that others do not know it does not make it a secret. There is in the word secret an implication of concealment, but concealment must be effective for information to remain secret. A second modification may help.

"James Bond knows many secrets."

Does he now know many things that are hidden from others or is he aware of many things that are hidden from him? Perhaps a further modification would help.

"James Bond knows every secret."

Everything that is hidden from others is known by James Bond. Or is it? Anything that is known cannot be a secret, can it? So, Mr Bond can only know information that other people do not, but that is nothing special. And he could not know absolutely every secret anyway. Whatever we wish to say about Mr Bond ends up being exactly the same as whatever could be said about anyone; there is information in his possession and information beyond his possession.

Language might seem like a quagmire, but it is essential to thinking communication. In evolving units of consciousness the universe has done something wonderful but it has also created its own worst nightmare. Remember the spare fundamentals in our SNPWC model?

DYNAMIC	PASSIVE
CHARGE fundamentals	Index fundamentals
8 – Qa (Quantum Deterioration)	
9 - SPARE	$M \div TV$
10 - SPARE	*SPARE*
CONSTANT fundamentals	*QEP fundamentals*
11 – Boltzman	*SPARE*
12	$QEP = Qc^2$

The purpose and function of these spare fundamentals must now surely be apparent. They are the starting place of consciousness and, true to the grand unified symmetry of the CSFC, possess an obvious symmetry across the spine of the contour; two dynamic fundamentals and two passive fundamentals. The extent to which they are filled determines the strength of the consciousness available from a particular SNPWC, so that the sum total of all spare fundamental values from all the SNPWCs within a mind creates the overall power of that mind. Since some variance will naturally occur between the many billions of SNPWCs within any particular unit of consciousness, it follows that the performance of a mind will vary from one isoflux to the next. And this combined with entropy can explain how a unit of consciousness may lose power towards the end of its life; degradation within specific fundamentals.

The position of the first spare fundamental on the dynamic side of the central spine, immediately after Quantum Deterioration, is ample support for the reflection of entropy in thought. And the slight dislocation between the relative positions of the dynamic spares and passive spares further complicates the functioning of consciousness; its dynamic interaction with the cosmos is slightly out of kilter with its passive archive. So, what precisely are the individual functions of each fundamental? On the dynamic side we have perception and logic and it is not coincidence that one spare cell is adjacent to the Boltzman cell, an energy constant. On the passive side are language and memory, fortuitously adjacent to Quantum Energy Potential and an index fundamental that divides mass by the product of time and volume.

It is with interest and amusement that one notices the breach between memory and language across the sets of index and QEP functions. It may never be possible to access the SNPWC world, for such small energy levels defy even the searching probes of nuclear science. Quantum Pulse Cascade may only ever be an elegant and harmonious theory, neither provable nor disprovable, that happens to neatly encompass all known science and the main areas of existence that lie beyond science.

The Final Structure

Assuming most non-derived values are 0.5, representing 50% capability at each level.

Section E: Selective Wave Interception

$Q^1 = 0.9$
$q_0 = 0.1$
$A^0 = 0.5$
$O^{sp} = \sqrt{A^0} = 0.707106781$

$$Q_i^a = \frac{Q^1 + q_0}{2} \times A^0 = \text{Quantum Interception Arc (Quinarcs)} = 0.25$$

The product of Quantum Interception Arc is 'view-eventuality'.

$$_vQ = Q_i^a \times O^{sp} \times Q = {_v}T{_v}V{_v}M = 0.25 \times 0.707106781 = 0.176776695$$

Section F: Multiple Wave Expansion

$ØT = 0.5Q_i^a$

$Q_E^A = Q_i^a - ØT = \text{Quantum Expansion Arc (Quexars)} = 0.125$

$$S = \text{success rate of } Q_E^A = \frac{Q_E^A}{Q} \times 100 = 12.5\%$$

Quantum Weirdness $QW = Q \times rf = 0.2$
Quantum Orthodoxy $QO = Q - QW = Q \times (1-rf) = 0.8$
$rf = \text{reduction factor as \% deviation from Quantum Orthodoxy} / 100 = 20$

$\beta = 1-rf = \text{broadcast efficiency and } R = \text{reception efficiency, substituting } Q_E^A \text{ for } Q \text{ in each calculation; say } R = 0.5$

Potential of transmitter $= \beta Q_E^A = 0.1$

And Potential of receiver $= RQ_E^A = 0.0625$

$$_pQ = \text{para-eventuality} = {_p}T{_p}V{_p}M = \beta Q_E^A \times RQ_E^A = 0.00625$$

Section G: Multiple Wave Conception

$$S_f \times S_e \times K = E_p$$
$$(i - a) \times (1 - pj) = P_p$$
$$R_e \times R_a \times A = C$$

Definitions of terms;

S_f = sensory faculty strength = 0.9

S_e = sensory experience strength = 0.8

K = knowledge strength = 0.5

E_p = total strength of elementary perception = 0.36

$(i - a)$ = impression less assumption = 0.9

$(1 - pj)$ = 1 less prejudice = 0.99

P_p = Personal preconception final strength = 0.891

R_e = strength of realisation = 0.8

R_a = strength of rationalisation = 0.5

A = strength of adaptability = 0.9

C = final strength of conception = 0.45

$Q_C^A = E_p \times P_p \times C$ = Quantum Concept Arc (Quonsar) = 0.200475

Quasi-eventuality, $_qQ$ with its factors $_qT \, _qV \, _qM$ is derived from the simple relationship

$_qQ = \, _vQ \, _pQ$ = 0.176776695 x 0.00625 = 0.001104854344

Section H: Selective Wave transmission

Focus eventuality ($_fQ$) and its components $_fT \, _fV \, _fM$ is the product of $_qQ$ multiplied by the % functionality of language (Л).

50% grasp of a language that is only 70% efficient risks producing only a 35% language functionality (Л).

In our example $_qQ$ = 0.001104854344 and Л = 0.35 so $_fQ$ = 0.000386699

Quantum Language Arc or Qualar - $Q_L^A = \, _fQ S$ where S = Stimulus potential

So where S = 0.99 Q_L^A = 0.000382832

A FINAL VALUE

If we look at the succession of arcs this is what we have:

Q_i^a = 0.25

Q_E^A = 0.125

Q_C^A = 0.200475

Q_L^A = 0.000382832

Whilst the eventuality values give the following succession;

$_vQ$ = 0.176776695

$_pQ$ = 0.00625

$_qQ$ = 0.001104854344

$_fQ$ = 0.000386699

Now more than ever the disintegration from reality to our comprehension of it is painfully apparent. The above input values have been in the order of 0.5 or even higher - perhaps as high as 0.99 - yet the outcome is so low; why is this so? Why does even the most powerful consciousness fall so far short of understanding the truth around it? How many of us really know what is happening, even on the smallest and most local of scales? Consider such a small quantum event as your own thumbnail. Do you know how many atoms it contains, how many particles are in those atoms, what the atomic isotope percentages are, what chemical combinations those atoms are arranged in and the precise micro-photon exchanges being conducted between them? No, you do not. Even if you are the world's number one biochemistry genius you still would not possess such accurate information. I rest my case.

The above figures are seemingly low, but in reality the input values even for the most powerful minds are going to be significantly lower and so the final outcome realistically should be on an even more minute scale. If our highest input value had been 0.1 the results would have really been shocking.

Q_i^a = 0.00505

O^{sp} = 0.071063352

$_vQ$ = 0.0003588699

$\emptyset T = 0.5Q_i^a$

Q_E^A = 0.002525

S = success rate of $Q_E^A = \dfrac{Q_E^A}{Q}$ x 100 = 0.2525%

Quantum Weirdness QW = Q x rf = 0.997475
Quantum Orthodoxy QO = Q - QW = Q x (1-rf) = 0.002525
R = 0.1

βQ_E^A = 6.375625 x 10^{-6}

RQ_E^A = 0.0002525

$_pQ$ = 1.609845313 x 10^{-9}

$$S_f \times S_e \times K = E_p$$
$$(i-a)x(1-pj) = P_p$$
$$R_e \times R_a \times A = C$$

Definitions of terms;

S_f = sensory faculty strength = 0.5

S_e = sensory experience strength = 0.5

K = knowledge strength = 0.1

E_p = total strength of elementary perception = 0.025

$(i-a)$ = impression less assumption = 0.5

$(1-pj)$ = 1 less prejudice = 0.5

P_p = Personal preconception final strength = 0.25

R_e = strength of realisation = 0.5

R_a = strength of rationalisation = 0.5

A = strength of adaptability = 0.1

C = final strength of conception = 0.025

$$Q_C^A = E_p \times P_p \times C = 0.00015625$$

$$_qQ = {_v}Q\,{_p}Q = 5.777250265 \times 10^{-13}$$

Л = 0.1 so $_fQ = 5.777250265 \times 10^{-14}$

$Q_L^A = {_f}QS$ (S = Stimulus potential); if S = 0.5 then $Q_L^A = 2.888625133 \times 10^{-14}$

This might seem extremely disappointing but in an infinite universe it allows some comprehension; the sobering thought is simply that it falls short of respectability.

EXPLAINING THE UNEXPLAINED

Ghosts

Patterns in the archive (PASSIVE) fundamentals within the SNPWCs of historic temporal edifices may well release some coded micro-photon energy carrying patterns that can be perceived by contemporary temporal edifices. The main requirement for the reception of such coded micro-photons is a collection of SNPWCs that is particularly compatible with the information being carried. There is nothing supernatural about this, just a collection of contours that have the right pattern to receive part of a historic broadcast.

Premonition

Just as SNPWC patterns from the past may resonate well with contemporary temporal edifices, so too might it be possible for locked-in future developments of a collection of SNPWCs to make themselves apparent to contemporary edifices that are sensitive to them. Again, the prerequisite is a pattern that can be hosted well by that of the receiver's averaged out SNPWC pattern. No mystical forces or jiggery-pokery, just everyday Quantum Pulse Cascade science.

Sixth Sense & Intuition

One particularly sensitive collection of SNPWCs may well be able to acquire greater than average information concerning another collection. If historic patterns can be intercepted and yet-to-materialise patterns previewed then it is also possible for contemporary patterns inaccessible to most people to occasionally find a host that can receive and interpret some data.

History and Memory

Given that the passive archive of every single SNPWC maintains precise information concerning the reactions participated in, with the addition of consciousness comes the potential to recall past configurations. This is what we refer to as memory. History is precisely coded within the PASSIVE fundamentals of every particle, but the academic subject we refer to as history is more about how units of consciousness have documented events around them, and is therefore largely based on the memories and perceptions of a succession of contemporary observers. We should be careful to avoid confusing such memories with solid facts that are only obtainable by decoding the SNPWC of all participating particles in an event.

Inequality

Two identical cars built in the same factory by the same workers on the same day, from the same consignment of materials, and then treated identically by their owners for ten years. One car develops repeated mechanical problems and rust, the other remains looking as good as the day it was built, and develops no mechanical faults. How can this be explained by Quantum Eventuality theory? Simple; the materials were not identical at SNPWC level. One set had a higher Quantum Weirdness factor.

Pulsewave Convergence

The psychologist Jung called this phenomenon synchronicity.
1. Two unrelated event objects with similar SNPWC configurations are drawn together.
2. Two related event objects acquire characteristics of each other's SNPWC, providing a link or bond even when they are apart.

Possible outcomes - ghosts (as we have already seen), telepathy and the following scenario: seeing or encountering something or someone familiar or similar to something or someone you are familiar with, even when you are separated from them.

I have personal experience of this. On 12/09/2005 I saw a woman almost the exact double of Lilia. She was the same height and build and was even dressed similarly with a similar handbag and similar hair. This was at the beginning of the week that Lilia's mother, Petrovna, was undergoing a hip operation. Lilia's double was outside the church where I had been praying for Petrovna. On the actual day of the operation, 15/09/2005, I encountered Lilia's patronymic name Vladimirovna.

TO SUMMARISE

It may never be possible to know a final answer for everything. In this book I have tried to advance a mathematical and scientific theory capable of resolving the main issues of science and the unknown. Even if by some stroke of luck I have hit the proverbial nail on its contentious head, infinity will always retain hidden and unfathomable depths.

The real question for any workable theory and its models is not so much "can we know everything?" but more "can we know anything?" I believe Quantum Eventuality gives a bigger and better yes than many other current postulations.

If there are any final words to say about my theory, my thoughts concerning reality, morality, philosophy and consciousness and the universe in general it is the following. The universe obeys strict behavioural laws and logic – its own kind of selfless morality; the universe is beyond us. What is beyond us obeys strict behavioural laws and logic, loyal to a selfless morality; we have no authority to act without at least some of this natural discipline. In order to be effective as participants in the universe we need as much as possible of this natural discipline; to act outside this universal morality is to be unnatural. The universe contains far more intelligence than mere humans can hope to command; in order to be as intelligent as possible we should be as close to the universe as possible.

Grahame Gordon Innes

14/03/2006